TK 5101 .D64 1986

Dordick, Herbert S., 1925-

Understanding modern
telecommunications

DATE DUE

NEW ENGLAND INSTITUTE
OF TECHNOLOGY
LEARNING RESOURCES CENTER

UNDERSTANDING MODERN TELECOMMUNICATIONS

McGraw-Hill Series in Mass Communication

Alan Wurtzel, *Consulting Editor*

Dordick: Understanding Modern Telecommunications
Gamble and Gamble: Introducing Mass Communication

UNDERSTANDING MODERN TELECOMMUNICATIONS

Herbert S. Dordick
Temple University

NEW ENGLAND INSTITUTE
OF TECHNOLOGY
LEARNING RESOURCES CENTER

McGRAW-HILL BOOK COMPANY
New York St. Louis San Francisco Auckland Bogotá
Hamburg Johannesburg London Madrid Mexico Montreal New Delhi
Panama Paris São Paulo Singapore Sydney Tokyo Toronto

This book was set in Optima by Publication Services.
The editors were Eric M. Munson and Kaye Pace;
the cover was designed by Laura Stover;
the cover illustrations were done by Laura Stover;
the production supervisor was Marietta Breitwieser.
Project supervision was done by Publication Services.
Halliday Lithograph Corporation was printer and binder.

**UNDERSTANDING MODERN
TELECOMMUNICATIONS**

Copyright © 1986 by McGraw-Hill, Inc. All rights reserved. Printed in the United States of America. Except as permitted under the United States Copyright Act of 1976, no part of this publication may be reproduced or distributed in any form or by any means, or stored in a data base or retrieval system, without the prior written permission of the publisher.

2 3 4 5 6 7 8 9 0   HALHAL   8 9 8 7 6

ISBN 0-07-017662-0

Library of Congress Cataloging in Publication Data

Dordick, Herbert S., date
  Understanding modern telecommunications.

  (McGraw-Hill series in mass communication)
  Includes bibliographies and index.
  1. Telecommunication. I. Title. II. Series.
TK5101.D64   1986   384   85-11382
ISBN 0-07-017662-0

For Ruth

CONTENTS

PREFACE   ix

PART 1   UNDERSTANDING YOUR FUTURE WORLDS

1   The Emerging Information Societies   3

2   Fundamental Technical Concepts   24

PART 2   THE TOOLS OF THE COMMUNICATIONS REVOLUTION

3   The Multitalented Semiconductor Chip   57

4   The Communications Satellite: Newton and Clarke Cooperate   75

5   Networks and Terminals: The Architecture of Telecommunications   108

PART 3   THE TELEPHONE: NO LONGER PLAIN OR OLD

6   Telephone Technology: Taking the Mystery Out of the Commonplace   125

7   Intelligent Telephones: Modernizing the Traditional POTS   148

8   Bits, Bytes, and Bauds over the Telephone: How to Talk to Your Computer by Telephone   166

PART 4 THE SECOND ERA OF BROADCASTING

9   Broadcast and Cable Systems: The Second Era
    of Broadcasting   187

10 Mobile Radio: The Fastest Growing Telecommunications
    Industry   222

PART 5 PREPARING THE WAY FOR THE INFORMATION SOCIETY

11 Publishing without Paper: Teletext and Videotex   247
12 Communicating with Computers   266
13 Network Information Services: A New Industry   283

PART 6 TRACKING THE FUTURE

14 Speculations about an Information Society   299

    GLOSSARY OF TERMS   310
    INDEX   317

PREFACE

Historians will determine if we have entered a new era, if we are, indeed, in the postindustrial society or the information society, or if we have experienced a communications revolution. Whatever the verdict, it is quite clear that telecommunications technologies have become very important in our lives.

The popular media have made these technologies seem mysterious. Indeed, the ubiquitous telephone has, for more than a century, been shrouded in mystery. We know very little about this extraordinarily complex system that touches us every day and only recently have we recognized how profoundly it has affected our lives. Computers, the microchip, satellites, and wired and microwave transmission techniques have made possible the expansive growth of television and have provided us with the ability to leap across time and space. Leisure has become electronic, government at all levels is increasingly computerized, and organizations cope valiantly with new roles for people in offices that are more and more mechanized.

This book seeks to demystify what has been made unnecessarily mysterious. We explain the telecommunications technologies, how they work, and what they can do for us. We examine their contributions to our lives today and how they are likely to impinge on our futures. We offer some order to your thinking about the telecommunications technologies at a time when the marketplace is in disorder and provide you with the tools to determine for yourself the importance of these technologies to your work and interests.

When I was a teenager, so many years ago, I read John Tyndall's *Sound*. This work fascinated me and had a great deal to do with my decision to pursue a career in science and technology. Many years later I discovered that John Tyndall was a major figure in the science of acoustics. How wonderful that this scientist should write a book that could entertain, inform, and shape the career direction of a teenager who had not the slightest knowledge of calculus or the laws of physics.

It is the too rare talent exhibited by the great British popularizers of science that I have sought to emulate in this work. I have tried to write a book on modern telecommunications in that tradition, to provide an explanation of the complex that is honest and not demeaning. You will tell me how well I have succeeded.

Writing is a lonely activity. Nevertheless, I must express my appreciation to my editors, Eric Munson and Kaye Pace, who patiently shepherded a host of reviewers to provide me with much valuable advice and encouragement. To these reviewers I express my appreciation for their care and often great kindness in preforming a difficult task. These include: Don Agostino, Indiana University; Donald Cushman, State University of Albany; Robert G. Gillespie, Gillespie, Folkner, and Associates; Richard Hezel, Syracuse University; Thomas Martin, Syracuse University; Mitchell L. Moss, New York University; John P. Witherspoon, San Diego State University; and Alan Wurtzel, ABC Television.

Herbert S. Dordick

UNDERSTANDING MODERN TELECOMMUNICATIONS

PART **ONE**

UNDERSTANDING YOUR FUTURE WORLDS

Historians have the luxury of looking backward. For those of us who tackle today, every day, we must engaged the future before it overwhelms us.

This is our task in Part 1.

This past thirty years have witnessed a rush to label eras, perhaps to beat the historians to their punch. Riesman made us members of a lonely crowd and McLuhan placed us in a global village. Peter Drucker argued that we live in an age of discontinuity and Leonard Bernstein serenaded us with the music of the age of anxiety. We have passed through the postcollective and the postliberal eras and are now in the third wave. We have experienced an electronic revolution as we have become a network nation shopping in a network marketplace. Now we are an information society, living in the information age.

Many information societies exist. The common denominator among them is their dependence on telecommunications. If any label is to be placed upon these many worlds, it may well be that of the era of telecommunications.

In Part 1 we explore and explain these many information societies. To do so requires that we understand the language and concepts of this world of telecommunications.

CHAPTER 1

THE EMERGING INFORMATION SOCIETIES

ON REVOLUTIONS

A wit with whom we are acquainted announced that the United States became an information society on February 1, 1984. This statement aroused numerous complaints from the academic community who found it just too simple. Our friendly wit argued that on that day 90 million or more telephone users received their first monthly phone bills after the divestiture of AT&T. One look at that bill, he said, made them instantly aware of the cost of information and communications. It raised their consciousness about information, and a society whose consciousness about this subject is raised is an information society.

No one actually woke up on the morning of February 2, 1984 and announced that he or she was now living in the information era. Nor did anyone in Manchester, England, in the eighteenth century announce to coworkers in the mill that the industrial revolution had dawned. Imagine what the others on the factory floor would have done to this poor soul. Nevertheless, these workers knew that their lives were much different from the lives of their parents and grandparents on farms and in villages mired in the hopelessness of poverty and hunger that pervaded the English countryside in the sixteenth and seventeenth centuries. In the factory towns of Europe the pull toward urbanization sparked by the industrial revolution and still under way in the burgeoning cities of Asia and South America brought with it soot, slums, crime, and, for many, opportunities unavailable in the countryside.

Just as the English factory workers in the eighteenth century recognized the changing nature of the world around them, so do we recognize that our world is changing. Telecommunications is a modern technology, a recent phenomenon whose impact throughout the world is only just now being experienced.

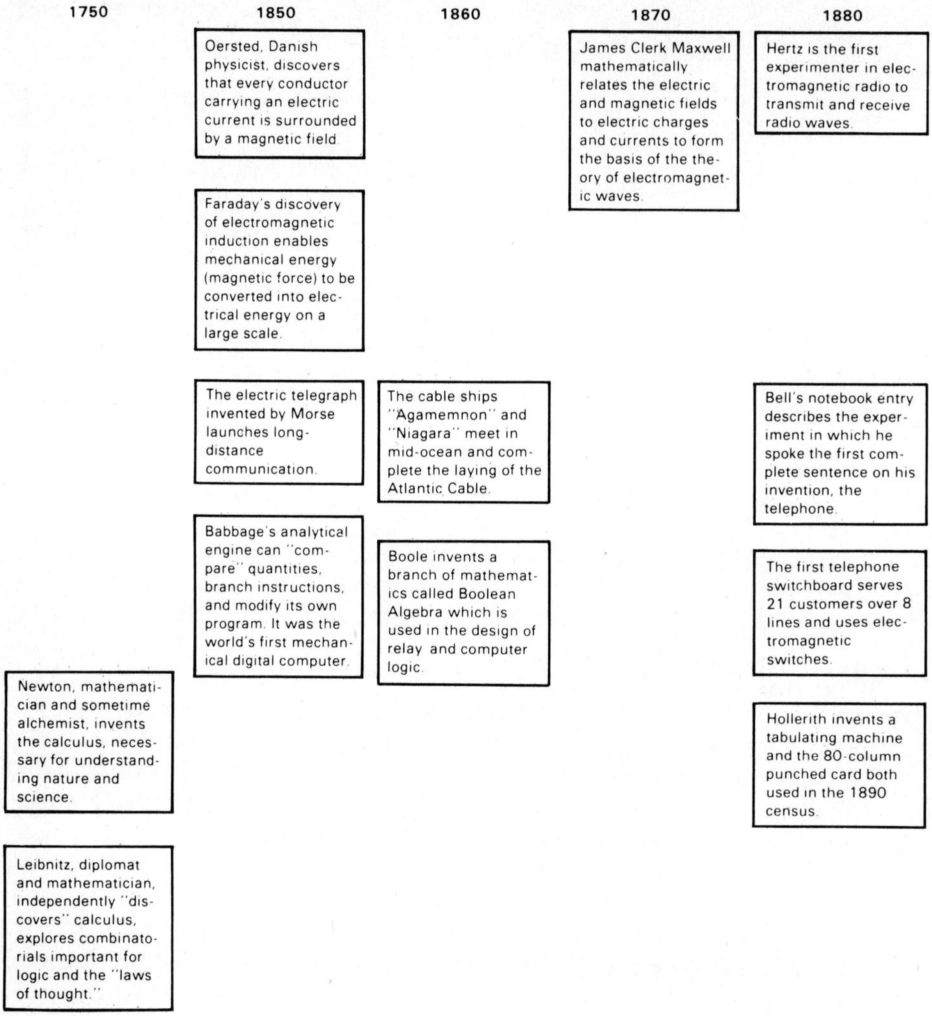

Timeline

1750

- Newton, mathematician and sometime alchemist, invents the calculus, necessary for understanding nature and science.
- Leibnitz, diplomat and mathematician, independently "discovers" calculus, explores combinatorials important for logic and the "laws of thought."

1850

- Oersted, Danish physicist, discovers that every conductor carrying an electric current is surrounded by a magnetic field.
- Faraday's discovery of electromagnetic induction enables mechanical energy (magnetic force) to be converted into electrical energy on a large scale.
- The electric telegraph invented by Morse launches long-distance communication.
- Babbage's analytical engine can "compare" quantities, branch instructions, and modify its own program. It was the world's first mechanical digital computer.

1860

- The cable ships "Agamemnon" and "Niagara" meet in mid-ocean and complete the laying of the Atlantic Cable.
- Boole invents a branch of mathematics called Boolean Algebra which is used in the design of relay and computer logic.

1870

- James Clerk Maxwell mathematically relates the electric and magnetic fields to electric charges and currents to form the basis of the theory of electromagnetic waves.

1880

- Hertz is the first experimenter in electromagnetic radio to transmit and receive radio waves.
- Bell's notebook entry describes the experiment in which he spoke the first complete sentence on his invention, the telephone.
- The first telephone switchboard serves 21 customers over 8 lines and uses electromagnetic switches.
- Hollerith invents a tabulating machine and the 80-column punched card both used in the 1890 census.

FIGURE 1-1
Major Figures in the Development of Telecommunications Technology. (Adapted from *A History of Telecommunications Technology,* prepared with the assistance of MIT and based upon a chart in the ITT Telecommunications Concepts Program, © 1980, for a symposium, "Communications in the 21st Century," sponsored by Phillip Morris Incorporated and used with permission.)

Even in the U.S.—this most information-based of information societies—there are rare folks who do not depend on television for their evening's entertainment and small communities just now installing their first telephones. The boom in telecommunications facilities throughout the world has been propelled by the satellite, and by faith in the benefits of a new industrial revolution based upon information rather than power, a faith that this information will offer a second

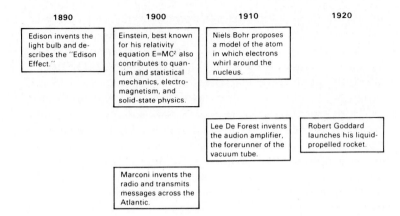

chance for the developing world to advance beyond the industrial revolution and create a more equitable world society.

The rapidity with which computers and electronic devices are invading the workplace and the home creates the impression that we are, indeed, experiencing a revolution—a communications revolution, or, more to the point, an information revolution. But innovations come slowly and revolutions rarely occur. Every invention has to wait years or even centuries before it is introduced into daily life. As Fernand Braudel puts it in his wonderful study of *The Structures of Everyday Life*,[1] "First comes the *inventio*, then, very much later, the application (*usurpatio*), society having attained the required degree of receptivity." Consider the steam engine; it was invented long before it launched the industrial revolution. (Or should we say, was launched by the industrial revolution?)

Today's communications "revolution" is not the result of the flowering of new science but of an explosion in applications of knowledge we have long possessed. The lineage of today's communications revolution is shown in Figure 1-1. The scientific basis for today's information economy and its communications technologies can be traced to Newton and Leibnitz in the eighteenth century, to Oersted, Faraday, and Babbage in the early nineteenth century, and to the work of Clerk Maxwell and Boole in the late nineteenth century and of Einstein and Bohr in the early twentieth century. Hertz followed in the footsteps of Faraday and Clerk Maxwell without whom Shannon might never have defined information as he did. In the United States Morse put Faraday's theories and experiments into practice in the telegraph, which in turn gave birth to a string of inventions, from the telephone to the computer. Von Neumann showed us how a computer should work, but only after Babbage tried, unsuccessfully as we shall see, to build one with mechanical gears. To paraphrase the words of Isaac Newton, modern telecommunications is built upon the shoulders of the seventeenth- and eighteenth-century experimenters who captured electricity from clouds and bottled its mysterious force in jars.

No new science is required for the full development of the information era. But technology alone does not create eras. We would be terribly shortsighted if we simply pointed to a new video game as evidence that we are experiencing a communication or information revolution and have entered a new age.

Revolutionary inventions are those that spark a stream of innovations leading to significant changes in society. Innovations leading to imbalances that either reinforce those already existing or create new ones in society, or leading to dangerous asymmetries between nations, will certainly be perceived as revolutionary. When innovations lead to a realignment of economic sectors, change the nature of human activities, alter the course of cultural development, and demand the restructure or invention of institutions, then we might argue that a revolution is under way and that when it is over we will have entered a new era. All of these dramatic events and more have been attributed to the communications and computer technologies and to modern telecommunications. These technologies are credited with giving birth to a new era—the information society.

THE IDEA OF AN INFORMATION SOCIETY

To Japan must go the credit for "inventing" the information society concept. In Japan the term "johoka shakai," or informationalized society, was coined around 1966 by a science, technology, and economics study group formed by the government to provide guidance to economic planners.[2] The phrase was analogous to the industrialized society the name historians had given to the industrial revolution.

An information society, they argued, is one in which there is an abundance in quantity and quality of information with all the necessary facilities for its distribution. (Economists of information, a relatively new breed, call this an abundance of information stocks and flows.) This information is easily, quickly, and efficiently distributed and converted into the form and purpose the user desires. Furthermore, in the information society information is universally available at affordable costs; in short, everyone can purchase the information they need because the price fits everyone's pocketbook.

The notion that the production and distribution of information is an economic activity of considerable value was first proposed and examined by the economist Fritz Machlup in 1962. His *The Production and Distribution of Knowledge in the United States*[3] may certainly have been the inspiration for the "johoka shakai." Machlup's pioneering work sought to *measure* the philosopher's concepts of knowledge and its use. Economists have long assumed that buyers and sellers have "knowledge of the markets," and that producers have "knowledge of the available technology." Machlup recognized (as Daniel Bell did after him) that the production of technical knowledge and the rise in productivity that may result from it had become increasingly important for economic growth. Yet we knew so very little about how much knowledge was available, how it was growing, or how its production and distribution affected the economy.

Machlup collected and categorized data about the communications media—printed matter, photography, the stage and cinema, the postal service, and, of course, broadcasting and the telephone. Nor did he forget the production and distribution of knowledge at conventions and meetings where the old-fashioned means of communication, face-to-face, took the place of the electronic media. He also examined the "information machines"; he measured the information produced by electronic computers, office information machines, and instruments for measurement, observation, and control.

It should come as no surprise that the Japanese married their "johoka shakai" to Machlup's economics. Here was an island society that, having outgrown its natural resources years ago, now found it necessary to uncover new ones. The Japanese people suggested that information and communication might very well open new economic frontiers for their country. The "smoke stack" industries—oil and steel—were not going to be around forever. Knowledge production and distribution quickly rose to the top of Japan's industrial planning agenda.

These two pioneering concepts—the Japanese "johoka shakai" and Machlup's information economy—gave birth to many notions of the information society, and, indeed, to many information societies.

THE MANY INFORMATION SOCIETIES

Pineapples and the Information Society

On a tiny island off the coast of Oahu in Hawaii pineapples are grown. Every week two farmers row to market in Honolulu in their small canoe and deliver ten dozen pineapples. They have delivered ten dozen pineapples every week for years, ever since their fathers and grandfathers discovered the market in Honolulu and what could be bought with the money they received from the sale of their pineapples.

One day one of the rowers became ill in Honolulu and could not return to the tiny island. When he recovered, he wandered around the marketplace and into the city away from the docks. There he made a great discovery; he found that the prices paid for pineapples change every now and then, depending upon the price of other foodstuffs and on what is happening in that far-off place called the United States. He discovered that all sorts of things, not sold on the docks, could be purchase in Honolulu—cloth and clothing, more shoes than he had ever dreamed of, and tools that might make his and his family's work easier. He also found that he could buy chemicals and other "potions" that were said to increase the number of pineapples that could be grown and, perhaps, make them bigger and more valuable.

When he returned to the island he told everyone what he had discovered. A village meeting was held and it was decided that the people on the island should grow more pineapples and that they should store some of them for delivery when the prices were higher. They also decided to send one of their brighter villagers to live in Honolulu to keep track of pineapple prices and other matters that might interest the pineapple islanders. Since they could buy much of their food in Honolulu and still have money left over to purchase dresses, shoes, television sets, and radios, more land was converted to growing pineapples. The islanders built bins to store their pineapples and purchased larger canoes. Two families moved to Honolulu and soon after, a telephone was installed on the island so they could tell the other villagers about pineapple prices and the goods they could buy with their profits.

Life on the tiny island changed radically. An information revolution had taken place. The island had become an information society!

Of course this is not a true story, at least not in the sense that the events occurred off the coast of Oahu. This hypothetical example demonstrates that high technology is not required to create an information society. A convincing argument can be made that this island was transformed into an information society, a society now conscious of the value of information.

Work and the Information Society

On our pineapple island in Hawaii some villagers no longer worked in the fields. Indeed, some villagers did not seem to work at all; they wandered among the storage bins with paper and pencil and talked on the telephone to the two former villagers in Honolulu who, in turn, spent most of their time talking to people in Honolulu, sending letters to the pineapple growers on the island, keeping track of the events in Honolulu, and reporting these goings on to the growers on the island.

Thus, these former villagers became the island's information workers. They "grew" information rather than pineapples. Their product helped the pineapple growers increase output and profits and assisted them in scheduling field work so they could spend more time doing other things (surfing perhaps?) and making better use of the profits they obtained from their fields. As the storage bins filled, the growers soon wanted to protect their pineapples until they were ready for market. A representative from Honolulu appeared to tell them about insurance and other forms of protection against fire and theft. Other representatives soon followed to tell them about the new chemicals available to grow better pineapples and, as in all information societies, accountants and lawyers offered their services. These representatives did not work in the fields; they, too, were the new breed of information workers.

Meanwhile, in the United States, Marc Porat at Stanford University had decided to expand on one of Machlup's chapters focusing on the notion that the evolution of a society from one that produces agricultural and manufactured goods to one that also produces and distributes knowledge requires a new kind of worker. (If only Porat knew of our mythical island, the evidence would have been quite clear. But myths do not reap academic rewards.)

In his report to the Department of Commerce entitled "The Information Economy: Definition and Measurement,"[4] Porat identified two kinds of information activities in the economy: primary information activities, or those in which the output is an information product or a service that is purchased for its own merit—what the accountants, lawyers, and agricultural advisors on our Hawaiian island were now offering to the villagers; and secondary information activities, or those performed in order to produce other products—the activities of the villagers who stayed in Honolulu and those taking inventory in the pineapple storage bins and warehouses.

Porat showed that the mix of jobs and products in the United States has been shifting toward information-related activities for more than forty years. His seven-volume report also showed that the information sector of the economy—that is, those activities that produced information as end products for sale and those activities that produced information for the production of what economists call intermediate products—was growing faster than the total economy.

Indeed, there are a great many more information activities taking place in the United States than on our mythical Hawaiian island. In fact, there are whole information industries about which some analysts are diligently compiling data. Table 1-1 shows one of these compilations to 1981.

By themselves, these compilations might not demonstrate that the United States is an information society. Porat has argued that what workers actually do is a better indication of an information society. He surprised everyone when he showed that by 1980 more people in the United States were engaged in information work than in any other kind of work; indeed, about 48 percent of the U.S. population was engaged in one form or another of information work, while only about 3 percent were in agriculture, slightly more than 20 percent in manufacturing, and about 30 percent in providing services. All of this is illustrated in Figure 1-2.

FIGURE 1-2
The U.S. Work Force: 1860–1980. (From *The Information Economy. Vol. 1: Definition & Measurement.* OT Special Publication 77-12(1), U.S. Dept. of Commerce.)

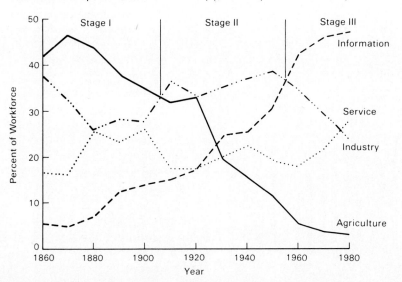

TABLE 1-1
REVENUES AND EXPENDITURES OF THE INFORMATION INDUSTRY IN THE U.S., 1970-1981 (IN BILLIONS)

Industry or institution	1970	1971	1972	1973	1974	1975	1976	1977	1978	1979	1980	1981
Communications												
Computer Software and Service Suppliers	$ 1.6	$ 1.8	$ 2.1	$ 2.6	$ 3.2	$ 3.8	$ 4.5	$ 5.3	$ 6.3	$ 7.5	$ 12.8	$ 15.3
Computer Systems Manufacturers	b	b	12.2+	14.4+	16.6+	18.8+	21.2+	23.8+	28.0+	31.2+	37.7+	42.6+
Electronic Components and Accessories	7.3	7.3	8.8	10.8	11.3	10.1	12.4	15.4	17.9	22.7	27.6	28.2
Mobile Radio Systems	1.9*	2.2*	2.4*	2.6*	2.9*	3.2*	3.5*	4.2*	5.0*	a	a	a
Satellite Carriers	0.0*	0.0*	0.1*	0.1*	0.1*	0.2*	0.2*	0.2*	0.3*	0.3*	0.4*	0.5*
Telegraph	0.4	0.4	0.4	0.5	0.5	0.5	0.5	0.6	0.6	0.6	1.2	1.4
Telephone	18.2	20.0	22.6	25.5	28.3	31.3	35.6	40.1	45.2	50.6	55.6	63.7
Terrestrial Common Carriers	0.0	0.0	0.0	0.0	0.0	0.0	0.1*	0.1	0.2	0.3	0.4	0.7
Media and Entertainment												
Advertising	1.4	1.4	1.6	1.7	2.0	2.1	2.5	2.8	3.5	4.0	3.7	4.6*
Broadcasting												
Radio	1.1	1.3	1.4	1.5	1.6	1.7	2.0	2.3	2.6	2.9	3.3	3.7*
TV	2.8	2.8	3.2	3.5	3.8	4.1	5.2	5.9	6.9	7.9	8.8	9.8
Book Publishing	2.4	2.7	2.9	3.1	3.3	3.5	4.0	4.9	5.4	5.5	6.1	6.9
Cable TV	0.3	0.3	0.4	0.5	0.5	0.9	1.0	1.2	1.5	1.8+	1.7	2.1*
News Wire Services	0.1+	0.1+	0.1+	0.1+	0.1+	0.1+	0.2+	0.2+	0.2+	0.2+	0.3+	0.3+
Motion Picture Distribution and Exhibition	1.2+	1.2+	1.4+	1.8	2.3	2.5	2.4	2.7	3.4	3.5	4.1	4.4
Newspaper Publishing	7.0	7.4	8.3	8.9	9.6	10.4	11.7	13.0	14.6	16.2	18.0	19.5*
Organized Sports, Arenas	1.1	1.2	1.2	1.2	1.4	1.4	1.6	1.8	1.9	2.1	2.3	a
Periodical Publishing	3.2	3.2	3.5	3.9	4.1	4.4	5.0	6.1	7.2	8.3	9.0	9.9*
Printing, Book, and Commercial	8.8	9.1	10.0	11.0	12.0	12.9	14.9	16.5	16.5	18.6	20.6	22.8*
Radio and TV Communications Equipment	9.3*	8.7*	9.1	9.7*	10.6*	11.9*	13.2*	14.9	16.9	19.6	23.8	27.8*
Theaters	0.1	0.1	0.1	0.1	0.1	0.1	0.1	0.2	0.2	0.3	0.4	0.5
Postal												
Postal Service	6.3	6.7	7.9	8.3	9.0	10.0	11.2	13.0	14.1	16.1	17.1	19.1
Private Delivery Services	0.8+	1.1+	1.3+	1.5+	1.7+	2.1+	2.3+	3.0+	3.5+	4.3+	5.2+	6.1+

Financial and Legal												
Banking and Credit	61.1+	68.9+	77.6	101.3	136.2	132.7	144.7	159.4	195.3	242.7	a	a
Brokerage Industries	40.6+	47.4+	55.3	61.0	64.1	69.1	80.6	59.4	68.2	92.2	a	a
Insurance	92.6+	103.5+	113.8	123.6	133.2	148.8	173.1	196.5	223.2	235.5	a	a
Legal Services	8.5	9.6	10.5	12.2	13.7	14.8	16.2	18.4	21.4	24.8	a	a
Miscellaneous Manufacturing												
Paper and Allied Products	9.5	9.8	11.0	12.9	17.0	16.2	18.9	20.2	21.9	25.4	28.7	35.2*
Photographic Equipment and Supplies	4.4	4.7	5.6	6.4	7.5	7.6	8.8	9.9	11.5	13.4	15.9	18.0*
Miscellaneous Services												
Business Consulting Services	0.9+	1.1+	1.1	1.5	1.7	1.8	2.2	2.6+	2.9+	4.7+	a	a
Business Information Services	0.8*	0.9*	1.0*	1.1*	1.1*	b	b	b	2.7*	b	b	4.1*
Marketing Research Services	b	b	b	b	b	0.3+	0.4+	0.4+	0.5+	0.6+	0.7+	0.8+
Total Revenue	$293.7	324.9	376.9	433.3	499.6	527.3	600.1	645.0	749.5	863.8	305.4	348.0
Government Expenditures												
Census Bureau	0.1	0.1	0.1	0.1	0.1	0.1	0.1	0.1	0.2		0.8	0.3
County Agents, Government	0.3	0.3	0.4	0.4	0.4	0.5	0.5	0.6	0.6*		0.7*	0.7*
Libraries	2.1	b	b	b	b	b	b	b	b	5.7*	6.6*	7.5*
National Intelligence Community	5.6*	5.4*	5.4*	5.7*	5.9*	6.3*	6.7*	7.4*	7.8*	8.3*	9.2*	11.0*
National Technical Information Service	0.0	0.0	0.0	0.0	0.0	0.0	0.0	0.0	0.0	0.0	0.0	0.0
Research and Development	15.3	15.5	16.5	16.8	17.4	19.0	20.8	24.0	26.5	29.0	31.6	a
Schooling	70.4	76.3	83.3	89.7	98.0	111.1	121.8	131.0	140.4	152.1	169.6*	181.3*
Social Security Administration	1.0	1.2	1.3	1.4	1.8	2.2	2.6	2.7	3.0	3.2	3.6	4.0
Total Expenditures	$94.8	98.8	107.0	114.1	123.6	139.1	152.5	165.7	184.1	193.4	222.1	204.8

Key:
* estimated; + lower bound; a not available as of January, 1983; b not available; c under $50 million annually
Copyright © 1984 Program on Information Resources Policy, Harvard University. Reprinted with permission from Benjamin M. Compaine "Shifting Boundaries in the Information Marketplace," in B. Compaine, ed., *Understanding New Media: Trends and Issues in Electronic Distribution of Information* (Cambridge, Mass.: Ballinger Publishing Co., 1984).

These statistics ought not to be too surprising. Government is an information activity and so is education. Broadcasting and publishing are information activities and writing this book is information work. Physicians are primarily information workers because, as Porat estimated, more than 80 percent of the activity in their offices is devoted to gathering information about their patients; perhaps less than 20 percent is spent doing something about their diagnosis such as setting the broken arm or removing the appendix. Dentists are also information workers, but to a lesser degree. Only about 20 percent of a dentist's time is given over to diagnosis; the remainder is taken up by the mechanical activities of working on teeth and dentures. The information industries listed in Table 1-1 gives some idea of the number of information workers in the U.S.

Somehow, being an information society has become a matter of national prestige. Using information workers and information industries as measures, other nations soon began their own analyses to determine if they, too, were information societies. Japan, of course, emerged as a very strong contender, but its figures were somewhat different from those in the United States. The Japanese estimated that only about 27 percent of their workers were engaged in information work with a slightly higher percentage still in manufacturing (see Figure 1-3). However, this determination is very much a matter of definition. For example, while Porat argues that a secretary is an information worker, the Japanese analysts point out that although a secretary may hold an information

FIGURE 1-3
The Japanese Work Force: 1955–1980. (Four-sector aggregation adapted from Kimio Uno, "The Role of Communications in Economic Development: The Japanese Experience," in *Communication Economics and Development*. M. Jussawalla & D.M. Lamberton, eds. Elmsford, N.Y.: Pergamon Press, 1982.)

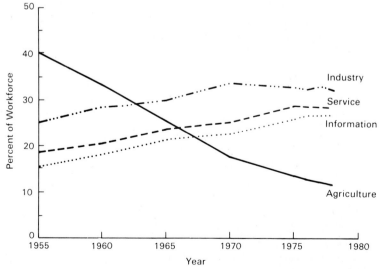

job, he or she is really a "blue collar" or manufacturing worker. The originator of the information the secretary processes is the Japanese information worker.

Clearly, defining information societies on the basis of information workers is tricky business and especially dangerous if politicians and economists intend to use the idea of an information society as a measure of a nation's quality of life.

The Quality of Life and the Information Society

If being higher on some information society scale is a measure of progress, then there must be some relationship between information and the quality of life. Most of us would agree that having a telephone is better than not having one and that without books life would be terribly empty. In many places in the world, radio is considered a necessity and television follows close behind. These are all instruments for the delivery of information. Information must, therefore, be important to the quality of our lives, but how to measure this impact is a challenge as yet unmet.

Japanese researchers have defined an information ratio—the ratio of the household expenditure for various kinds of information-related activities to total household expenditures. It is difficult to determine these expenditures; however, they might include the amounts a family spends on newspapers; books; radios; television sets; magazines about radio, television, fishing, hunting, and other special interests; education; entertainment, including cinemas, lectures, museums, and so on. Most people do not keep such close track of their household accounts, so the Japanese approximated this ratio by subtracting all major expenses such as food, clothing, shelter, and other clearly non-information-related expenditures from the total household budget. The amount left over was labeled "expenditures for information." (Those who remember their economics will note the similarity to the famous Engels' ratio—the ratio of expenditures for food to total household expenditures.) Figure 1-4 shows how various developed nations rank on the information ratio to per capita income scale.

For an interesting exercise in speculation, compare what you know or have heard about life in the countries listed here with their ranking on the information ratio scale. Would you say that the higher a country appears on the scale the better its quality of life?

Rather than measure expenditures for information activities and equipment why not measure information use? Indices have been developed in the U.S. and Japan to track the flow of information. Counting words transmitted and words attended to or used might give us a better idea of the nature of an information society.

Consider, however, the work entailed in counting words in telephone conversations and on radio and television broadcasts or transmitted by pictures, and converting these counts into words per minute. The Japanese have had the patience to undertake this task. They found that the supply of information

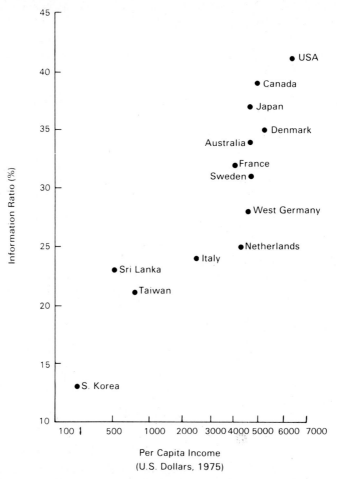

FIGURE 1-4
How Nations Rank as Information Societies. (Adapted from Annual Report, Research Institute of Technology and Economics, Tokyo, Japan)

delivered by various media has risen over the past decade and a half, with the exception of the cinemas, perhaps indicating a reduction in the number of films produced as television broadcasting has increased (see Table 1-2).

Let's examine how much of this supply of information is actually consumed. For the telephone, telegraph, and other point-to-point media—the personal or small media as compared to the mass media such as radio and television—the number of words supplied is the same as the number of words consumed. This is not the case for the mass media, as shown in Table 1-3. The supply of information or words far exceeds the consumption!

TABLE 1-2
THE SUPPLY OF INFORMATION IN JAPAN: 1960–1975

Media	Year				Growth rate per annum (%)
	1960	1965	1970	1975	
Telephone	31,250	42,193	89,853	124,389	9.6
Telegram	5	5	5	6	1.2
Telex	672	1,095	2,721	2,672	9.6
Facsimile	23	72	174	603	24.3
Data Transmission		35	1,707	67,360	65.6
Mail	28,380	41,070	48,760	52,090	4.1
Direct Mail	98,300	142,800	168,900	177,000	4.0
Newspaper Pre-printed Inserts	—	—	—	227,000	—
Radio	21,150,000	24,430,000	44,920,000	68,710,000	8.2
Television	6,536,000	23,478,000	35,375,000	84,585,000	18.6
CATV	110,500	383,400	728,000	1,282,100	17.8
Newspapers	3,919,000	5,364,000	9,910,000	10,835,000	7.0
Books	121,400	220,300	374,700	483,300	9.6
Magazines	514,100	660,500	1,133,400	1,400,900	6.9
Telephone Directory	59,500	99,000	282,000	392,000	13.4
Movie	1,196	746	521	392	-7.2
Education	70,500	77,200	68,600	72,600	0.2
Total	32,640,826	54,940,416	93,104,341	168,412,412	11.6

Unit: 10^8 words.
Source: Ithiel de Sola Pool, Hiroshi Inose, Nozomu Takasaki, and Roger Hurwitz: *Communications Flows: A Census in the United States and Japan*, (Tokyo: University of Tokyo Press, 1984). Reproduced by permission.

At MIT, researchers have replicated this Japanese work on information supply and consumption for the United States.[5] They found that from 1960 to 1977 words made available to Americans through the various media grew at a rate of 8.9 percent per year or more than double the 3.7 percent per year growth rate for the gross national product. However, even taking into account population growth, the words actually consumed or attended to grew at a rate of only 1.2 percent per year. It would seem that in the U.S. as well the supply of information continues to far exceed the consumption.

How do we interpret this phenomenon? Do we say that in both the United States and Japan, considered by any definition the two most advanced information societies, information resources are being used highly inefficiently? Or ought we to argue that the oversupply of information represents the workings of democratic societies, societies that traditionally offer their citizens a full marketplace of ideas from which to choose? Thomas Jefferson would probably

TABLE 1-3
THE CONSUMPTION OF INFORMATION IN JAPAN: 1960-1975

Media	Year				Growth rate per anuum (%)
	1960	1965	1970	1975	
Telephone	31,250	42,193	89,853	124,389	9.6
Telegram	5	5	5	6	1.2
Telex	672	1,095	2,721	2,672	9.6
Facsimile	23	72	174	603	24.3
Data Transmission		35	1,707	67,360	65.6
Mail	28,380	41,070	48,760	52,090	4.1
Direct Mail	55,500	80,700	95,400	100,000	4.0
Newspaper Pre-printed Inserts	—	—	—	269,600	—
Radio	1,859,000	508,000	543,000	749,000	-5.9
Television	1,215,000	3,993,000	4,474,000	5,337,000	10.4
CATV	16,410	75,720	106,830	214,160	18.7
Newspapers	1,027,600	1,138,000	1,119,900	1,263,200	1.4
Books	223,000	248,000	272,000	322,000	2.5
Magazines	238,000	263,000	289,000	345,000	2.5
Telephone Directory	244	272	290	304	1.5
Movie	82,700	30,400	20,800	14,200	-11.1
Education	885,000	923,000	801,200	863,300	-0.2
Total	5,662,179	7,344,562	7,865,640	9,724,884	3.7

Unit: 10^8 words.

Source: Ithile de Sola Pool, Hiroshi Inose, Nozomu Takasaki, and Roger Hurwitz, *Communications Flows: A Census in the United States and Japan*, (Tokyo: University of Tokyo Press, 1984). Reproduced by permission.

agree with the second of these two interpretations; he argued, if you remember, that an informed electorate is the best guarantee of a democratic society.

The Network Marketplace and the Information Society

Much has been written about the marriage of computers and telecommunications, or the convergence of computing and communicating. In this book, we shall see how this phenomenon is taking place, for it is this convergence that is changing the shape of the telecommunications technologies.

In the United States and increasingly in many of the highly industrialized nations of the world, a new industry has emerged from the marriage of computers and telecommunications. This industry permits users to interact directly with one or more computers in the form of distributed information systems which may be found in banks, shops, airline terminals, libraries, and schools.

Some analysts have suggested that the potential for the delivery of services via a network (sometimes called network information services) could virtually transform not only the information and communications activities of a society, but the very nature of the society itself.[6] They suggest that new services will be offered, such as remote shopping, news on demand, electronic mail or message delivery, remote medical consultation and diagnosis, electronic banking, remote and interactive education and training, and, conceivably, "remote" work situations.

Network information services will connect the needs and resources of users to the capabilities and services of producers and thus facilitate transactions between the two. All of the usual services of a marketplace can be offered within a large information network. Products and services can be advertised; buyers and sellers located; and ordering, billing, and delivery of services facilitated. All manner of transactions can be consummated, including wholesale, retail, brokering, and mass distribution. Indeed, the entire range of products and services offered to consumers by business, industry, and government can be viewed as a network marketplace.

Many societies throughout the world, and particularly the United States, seem to be ready for this transformation. Driven by human nature and abetted by technological opportunities, we seem to be moving toward an isolated way of life. Some studies of the potential for network information services in the United States certainly support this conclusion. Look at Table 1-4 and note the high percentages of people who seem willing, if not eager, to do their shopping, banking, and learning, on a network.

The information technologies, communicating and computing, have not only reached into the office, but have also created a most unusual electronic

TABLE 1-4
HOW PEOPLE REPORT THEY WILL LIVE ON THE INFORMATION NETWORK

Service	1985		1980		1995	
	% Households	$/mo	%Households	$/mo	%Households	$/mo
Banking	10	3.00	30	4.47	50	5.96
Shopping	5	6.00	15	8.54	30	11.42
Electronic Mail	10	1.50	20	1.95	30	2.00
Security	10	22.00	20	29.50	40	39.40
Education	10	47.70	15	55.20	20	64.00
Entertainment	25	26.70	40	35.60	50	47.60
Games	10	10.60	20	14.25	30	19.00
Library Serv.	10	12.00	15	15.00	20	20.00
Gen. Information	5	13.40	10	18.00	25	24.00
Population	234,000,000		245,000,000		258,000,000	
No. of Households	88,640,000		94,231,000		97,357,000	

(Based upon a 15-city survey, 1979-1982, by the author).

environment in the home, what we could call the "networked home." The ubiquitous telephone reaches into more than 96 percent of all of the nation's households and is certain, as we shall see in subsequent chapters, to become a more sophisticated information instrument, its simple keypad expanded to typewriter format and married to a cathode ray tube that is likely to be the television set. By the mid-1990s at least half of all television households in the United States will be "on cable."

Few if any households in the nation are not now within reach of a television and radio broadcast signal. We shall learn that the empty "spaces" in that signal can be used for electronic publishing, or teletext. Direct broadcast satellites beaming signals into homes unreached by terrestrial broadcast networks will deliver information in various modes in the very near future and enlarge the range of information offered to homes that are reached by broadcast networks.

An important by-product of the convergence of communications and computers has been the discovery that the electromagnetic spectrum can be utilized much more efficiently if we are clever in dividing up the portions we use. In subsequent chapters we shall examine how this valuable resource can be conserved and why new communications technologies such as cellular radio and specialized microwave services are creating additional networks for consumers and business. With the telephone, twisted pair cable, or glass fiber, we can network the home, the office, and, indeed, the nation.

Do these networks that crisscross the country—wire and broadcast, terrestrial and space, copper and glass fiber—provide the infrastructure for yet another information society? Are these networks the prerequisites for any information society?

MODERN TELECOMMUNICATIONS AND THE INFORMATION SOCIETY

Just as the hammer and chisel are the tools of the stone cutter and the plane and saw those of the carpenter, so are the telecommunications technologies the tools of the information worker. The pineapple growers on our mythical Hawaiian island quickly learned the value of the telephone and soon became quite dependent on the weekly reports they received from Honolulu. They discovered, as today's managers are discovering, that the telephone may, indeed, be their most valuable information tool. So commonplace has the telephone become in our lives, we often overlook its importance.

The telecommunications technologies make information societies possible. Many forces work toward the development of information societies: economic, political, social, human behavior, and more. Telecommunications provides the means by which information is communicated and thereby becomes of value to a society.

The decades following World War II have witnessed a veritable explosion in communications technology. It seems as if the war stimulated a need throughout the entire world to "reach out and touch someone." Telephone subscribership in the United States almost doubled during the period from 1945 to 1975 and oceans were crossed with more transoceanic cables than had been laid in the previous century. Our friendly POTS, plain old telephone services, learned the new language of bits, bytes, and bauds and now leaps across continents through outer space rather than over wires.

Television graduated from a laboratory curiosity to a global phenomenon researched by numerous theoreticians who seek to understand the secret of its power to "glue" millions to the blue haze of the picture tube. Television, too, is learning a new language. Its messages are no longer limited to moving images and sound transmitted over scarce and jealously guarded electromagnetic waves. Television, too, leaps across continents through space delivery data and news. With the introduction of cable TV, there seem to be no limit to the number of delivery channels that can be accessed.

Radio was to have succumbed to the overwhelming power of the picture, but instead it has found new roles in the information society. Radio allows peripatetic modern man to become independent of time and space, to work while traveling, swimming, and playing, indeed, to never be "out of touch."

Now that the telephone can speak the language of computers, the computer itself has become a communications tool. Data communications allows for computers to "talk" to other computers and for people to talk to one another even though separated by time as well as space. We will see in the chapters that follow how electronic mail may rid us of "telephone tag," that frustrating game we are often forced to play across time zones.

It is rather remarkable that we never seem to give up one medium entirely in favor of the latest, "more modern" one. It is the nature of communications media that one builds upon another. The 145-year-old telegraph is still with us in both its original form as well as in the form of the modern Telex or switched telegraph. Unlike the horse and buggy which was replaced by the automobile (except in such picturesque places as the Amish country of Pennsylvania), the telegraph has not disappeared. Rather, it continues to play a most important role in many nations of the world and is integrated into some of the most modern international computer-communications networks. Networks that arrange for your 'round the world air travel and hotel reservations can communicate equally well with a telegraph that sends only five words per minute and a computer that sends thousands, even millions, of words per second.

The explosion in telecommunications technologies does not belong solely to the U.S. or to the most developed nations of the world. It is a global happening. As never before, the nations of the world will be interdependent on and interrelated with one another through a worldwide telecommunications network. Modern telecommunications may yet create the one world dreamers have envisioned for years.

Telecommunications has moved to the top of the political agenda in many nations. New technology is breaking up the monopolies that controlled telecommunications. AT&T, the British Post and Telegraph, and the French Telecommunications monopolies are currently under competitive challenges. Even the powerful Ministry of Posts and Telecommunications in Japan has come under fire from the country's trade ministry to more rapidly open its doors to competition. The growing importance of telecommunications throughout the world is the result of three important factors:

1 Many nations have come to the realization that they cannot compete in the development of information industries without adequate telecommunications technology.

2 Because of the convergence of the computer and communications technologies these industries see possibility for growth in each other's territories.

3 Multinational corporate operations require worldwide, round-the-clock communications in the form of global networks for banks, air traffic control, travel reservations, news, and trade.

THE IMPORTANCE OF UNDERSTANDING MODERN TELECOMMUNICATIONS

In less than three decades, electronic communications and information technologies have restructured the nature of business, industry, and the family throughout much of the world. In the United States, where these technologies represent the leading edge of the nation's industrial and political strengths, significant changes in the shape of industry and the nature of work have begun to emerge.

Telecommunications is radically altering our lives. In the home, on the farm, in the office, and on the road, we shall be doing our work and enjoying our leisure time in new ways.

We can expect telecommunications technology to invade every aspect of our lives; transmission costs will fall to nearly zero as we apply high-speed computing to communications. Terminals and other intelligent devices will become smaller and cheaper and engineers will design new information systems based on the premise that transmission is essentially free. Channels of communication will be plentiful, and to Dick Tracy's wrist watch radio will be added a television receiver and transmitter with a computer thrown in for good measure. Every man, woman, and child in the United States will have an opportunity to communicate with every other man, woman, and child on earth. In the not-too-distant future, every working American will be able to have at least two hours a day of uninterrupted computer time, with time left over to appease the more talkative among us.

This book seeks to provide an understanding of the nature of the telecommunications technologies. We wish to demystify what the popular press has so often mystified; to allow you to see beyond the promotion and hype of the information age. We hope to provide the tools by which you will be able to:

Understand communications and the information revolution.
Motivate yourself to study and learn about them,

Explain how the technologies work,
Understand what they can do, and
Know what the future has in store.

THE ROADMAP FOR THIS BOOK

Following this brief visit to the many information societies, we shall examine the fundamental concepts necessary to understand modern telecommunications. This examination will teach us about the tools of the communications revolution: the semiconductor chip, the satellite, and the somewhat less glamorous but nonetheless necessary networks and terminals which go into the architecture of our telecommunications systems.

In Part 3, we will demystify that old standby, the telephone, and show how it is being modernized and becoming an essential and, indeed some might say, sufficient tool for an information society. We will show how and why the "digital" telegraph has finally converted the more familiar analog telephone.

The wireless networks surrounding us that both support and compete with wired networks will be examined in Part 4. We will demonstrate how these networks are converging just as computing and communications have converged.

Telecommunications systems are the pathways for the delivery of information. In Part 5 we show how the tools described in Part 2 are making possible the efficient delivery of information. Here, you will be introduced to data communications, a topic deserving its own text.

We avoid the usual temptation to forecast technology in our final chapter. Instead, we speculate on the future of information societies by examining the nature of human behavior in historical terms. What better way to forecast the future than to look at the past?

We *do not* wish to make technological determinists or technocrats of our readers. Technology must have a human face, otherwise it will not serve, for no matter what the technology buffs argue, culture *is* more powerful than technology. In any study of modern telecommunications it is wise to always remember that there is a tradeoff between efficiency and equity and that the social and economic consequences of new technologies cannot always be anticipated. How you fare in this tradeoff depends upon your values.

The United States and much of the world faces dilemmas perhaps never before encountered so universally. We face uncertainties of energy costs and supply, and of raw materials in general. We live in an interdependent world where populations continue to increase while economies struggle along or threaten to fail entirely. As we learn more about the environment, we recognize new threats from pollution. To these concerns must be added the special needs of the world's least developed nations, and of the poor in even the most highly developed countries including the United States. Neither one is going to be left behind as the world moves into a new information era; developing peoples are determined to reap their rightful share of the rewards that come with the new technologies.

There are serious consequences for any society that chooses to enter the information era and adopt the technologies we shall be discussing. As you study modern telecommunications think of these consequences and how your own work will affect them and be affected by them:

- Modern telecommunications and the associated information technologies create the possibility of enormous increases in productivity in all sectors of the economy, including the home, factory, and office. With parallel developments in biochemistry and biochemical engineering promising to further increase agricultural production without increasing the need for agricultural workers, will we be faced with prospects of long-term and severe structural unemployment?

- Increasing use of computer-communications networks threaten the individual's control over personal privacy, perhaps because the complexities of these new technologies are inadequately understood. Perceptions are distorted and a feeling of personal powerlessness often results. Furthermore, there is a sense that access to information is increasingly restricted and that information will become an economic good with a price tag that is out of the reach of many. Will a new gap be created between the information rich and the information poor?

- Traditional educational methods and institutions may not be prepared to deal with the demands for new and frequently changing skills brought about by an information society. Should we alter traditional relationships between universities and industry and government?

- Intelligent machines and equally intelligent telecommunications systems and equipment may drastically alter the shape of manufacturing and assembly practices in the factory and the office. How will relationships between management and workers be changed? Will reduction in the office work force as has already been reported lead to important changes in the industry itself? Is the idea that manufacturing industries are becoming service industries and service industries are becoming increasingly automated more fact than slogan?

- Will the use of modern telecommunications by government and business further centralize authority and increase the power of the technocrats? Will personal relationships be destroyed as more and more people work and live in the network marketplace?

- In the long run will the trend be toward shorter working hours, less physical work, and indeed, less work? Will we need to redefine the terms work and unemployment?

Uncertainty has always been the curse of those who reflect on the nature of rapidly changing societies. Only later historians have the luxury, it seems, to claim with certainty that an age has dawned or a revolution has occurred. We know that the industrial revolution created inequities throughout the world. Some nations counted themselves among the beneficiaries while others saw themselves falling further behind, gaining little from the experience. We hope to do better this time.

REFERENCES

1 Braudel, Fernand, *The Structure of Everyday Life: The Limits of the Possible* (New York: Harper and Row, 1981), pp. 335–337.
2 Ito, Youichi, "The 'Johoka Shakai' Approach to the Study of Communications in Japan," in G. C. Wilhoit and H. de Bock, eds., *Mass Communications Review Yearbook*, Volume 2 (Beverly Hills, CA: Sage, 1981), pp. 671–698.
3 Machlup, Fritz, *The Production and Distribution of Knowledge in the United States* (Princeton, NJ: Princeton University Press, 1962), pp. 3–10, pp. 348–361.
4 Porat, Marc U., *The Information Economy: Definition and Measurement*, Volume 1 (Washington, DC: OT Special Publication, 77-12(1), 1977), pp. 15–21.
5 de Sola Pool, Ithiel, "Tracking the Flow of Information," *Science*, Volume 221, No. 4611, August 12, 1983.
6 Dordick, Herbert S., Helen G. Bradley, and Burt Nanus, *The Emerging Network Marketplace* 2nd ed (Norwood, NJ: Ablex Publishing Corporation), forthcoming, pp. 3–8.

ADDITIONAL READINGS

The literature heralding the birth of the information society ranges from the popular in weekly magazines to the soul-searching in studies by sociologists, historians, and philosophers. This selection will provide a glimpse of what is being said by writers from both schools.

Bell, Daniel, *The Coming of Postindustrial Society: A Venture in Social Forecasting* (New York: Basic Books, 1973). This monumental work launched intellectual discussions of information as well as postindustrial societies.
Drucker, Peter F., *The Age of Discontinuity* (New York: Harper and Row, 1973). Drucker interprets the postindustrial world for the manager.
Edelstein, Alex, John E. Bowers, and Sheldon M. Harsel, eds., *Information Societies: Comparing the Japanese and American Experiences* (Seattle, WA: International Communications Center, 1979). An early attempt to bring together observers from the two leading information societies.
Machlup, Fritz, *The Production and Distribution of Knowledge in the United States* (Princeton, NJ: Princeton University Press, 1962). If the information society is to have a founding father, it must be Machlup.
Masuda, Yoneji, *The Information Society: A Postindustrial Society* (Tokyo: Institute for the Information Society, 1981). This is a highly creative philosophical view of a concept about the information society.
Naisbit, John, *Megatrends: Ten New Directions Transforming Our Lives* (New York: Warner Books, 1982).
Nora, Simon, and Alain Minc, *The Computerization of Society: A Report to the President of France* (Cambridge, MA: MIT Press, 1980). This document was prepared for the president of France and is a model of public policy statements. It serves as a road map for the French information technology development.
Toffler, Alvin, *The Third Wave* (New York: William Morrow & Co., 1980).
Williams, Frederick, *The Communications Revolution* (Beverly Hills, CA: Sage, 1982). Everyone's guide to how communications technology might impact our lives.

CHAPTER 2

FUNDAMENTAL TECHNICAL CONCEPTS

The purpose of this chapter is to review some fundamental concepts in telecommunications and to define many of the terms used in this book. We begin by examining the basic elements of any communication system. Next, we discuss the various ways in which electrical signals are used to convey information and explain the distinction between analog and digital systems. We identify the physical limitations which constrain telecommunication systems and look at various means to overcome them. We review how several different signals can be sent over the same wire, cable, or spectrum space. Finally, we try to understand just what is the "information" that telecommunication systems are designed to carry.

To be an intelligent user and planner of communication systems, you must be able to "speak the language." This chapter is designed to accomplish that.

COMMUNICATION SYSTEMS AND NETWORKS

Communication has taken place in many ways throughout history. In early times, sign language was probably used, along with a great deal of very expressive body language. This form of communication has not disappeared altogether; for the deaf or those with severely impaired hearing, sign language remains a necessity and, if one is to believe some popular writers, body language is still effective for special purposes.

With the evolution of language, face-to-face speaking or shouting, as is often done in remote Japanese villages and small Alaskan towns, has essentially replaced sign language. Swiss yodeling is, perhaps, a more musical form of voice communication. Two tin cans connected by a string may have been your encounter with mediated human communications—communications via the vibrations of a string and thin tin.

With the invention of language, the conversion of recognized words into signs, smoke signals, or drum beats constituted a code or representation of the sounds that make up words. Indeed, one might even think of language as a code for thoughts and feelings.

We are concerned, however, with telecommunication systems, the different forms they take, and the different technologies they use.

All communication systems, no matter how sophisticated, can be perceived as having the same elements:

- An information *source*—you are an information source.
- An information *encoder*—your vocal tract takes ideas and encodes them into speech or vibrations from the tin can on a string; your vocal chords create changes in the air forming waves, peaks, and valleys, or vibrations.
- A communicating *channel*—the piece of string.
- A *receiver*—the other tin can which converts the vibrations into sounds.
- A *decoder*—your ear and brain, which translates the sounds into words and ideas.
- A *recipient*—the person to whom you are talking.

In today's electronic world, many different technologies and systems exist for each of these functions. The study of telecommunications is concerned with what they are, how they work, what they are good for, and how to choose among them for the task at hand. For example, the *source* might be an orchestra on the stage at Carnegie Hall; the *encoder* one or more television cameras encoding the scene for transmission over a broadcast *channel* to television antennae on rooftops which are linked to the *receivers*, the television sets in living rooms, where the signals are *decoded* by the picture tube and the set's electronics for delivery to the intelligent *recipient*, the viewer-listener.

Some forms of communication are virtually instantaneous, the only delay being the physical limitations of the speed of sound or light. Electrons travel at the speed of light, delayed only by the nature of the material through which they travel. Telephone conversations, live television either over-the-air (broadcast) or over-the-wire (cablecast), African tom-toms, and the semaphores still used by passing ships at sea all provide instantaneous, or real-time, communications.

In other forms of communication the message is first recorded and stored for some time before the recipient gets it. Books, films, paintings, recordings, Telex messages, messages from computer to computer, and the garbled "notes" on telephone answering machines are all examples of delayed communication.

So far we have talked about two persons communicating and an orchestra performing for many people. These types of communications are called point-to-point and point-to-many-points, respectively. Note that point-to-many-points communication encompasses broadcasting.

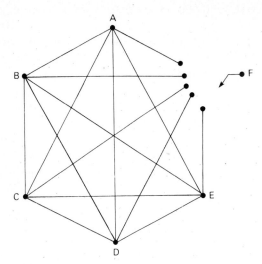

FIGURE 2-1
A Point-to-Point Network with Five Switches for Each Party.

When many people wish to communicate with each other point-to-point simultaneously, networks must be provided. One kind of network has a separate channel between each information source and recipient as shown in Figure 2-1. Each person has a device or switch to use in selecting from among all possible persons on the network with whom they might connect. Each person must have as many switches as there are people on the network, less one. Thus if there are ten people with whom you may wish, at one time or another, to talk individually, you will need nine switches to connect yourself, one at a time, to these ten people. That means in this network of ten people there would have to be ninety switches.

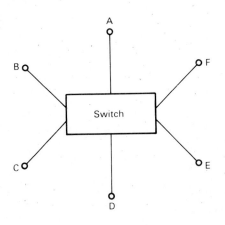

FIGURE 2-2
A Centrally Switched Network with One Reliable Switch for All Parties.

An alternative that was adopted very early in telecommunications is a centrally switched network as shown in Figure 2-2. This network uses fewer and shorter wires or cables and requires only one switch which must be very reliable and is often quite complex. In our study of telecommunications we shall be concerned with networks, as well as point-to-point communication.

TELECOMMUNICATION SYSTEMS

Telecommunication systems use electricity and electromagnetism to transmit messages or information. Consider the simple telegraph—it works by sending electric current over a wire. The telegrapher's sending key alternately opens and closes the connection between the sender and the receiver, or circuit, allowing current to flow as shown in Figure 2-3(a). At the receiving end, the current flows through an electromagnet which pulls on the receiver causing it to click.

Since the only language or symbols capable of being transmitted this way are electrical impulses, a code is required. The Morse Code uses dots and dashes (a dash is three times longer than a dot) to represent letters.

In practice, when you apply voltage to a line the current does not begin to flow immediately. Rather, it builds up over time and at a rate that depends on the nature of the material (the wire or cable) through which it is passing.

FIGURE 2-3
Signals Being Distorted During Transmission.

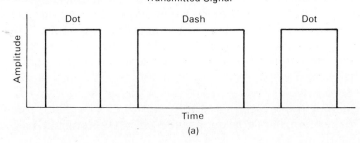

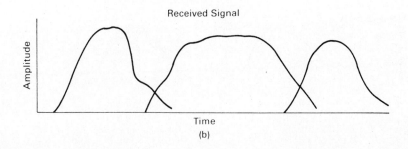

Consequently, there is a difference between the transmitted signal (Figure 2-3(a)) and the received signal (Figure 2-3(b)). This could mean that the information is not received as it was transmitted; there is *distortion* present. The difference between signals sent and signals received is shown again in Figure 2-4; the neat square dots and dashes of our telegraph signal in Figure 2-4(a) become curved when they are received (Figure 2-4(b)).

Signal distortion due to the characteristics of the transmission system is not the only problem facing designers of telecommunication systems; electrical noise is always present as shown in Figure 2-4(c). Static on your radio or snow and ghosts on your television screen are examples of this noise. The noise may be manmade as that generated by fluorescent lights or a computer when it is moving data around, or it may be caused by nature in the atmosphere (lightning for example) and in space when the sun is discharging flares. Scientists also have identified cosmic noise emitted from suns or stars light years away, noise that, as you will see in later chapters, has an effect on our ability to communicate with satellites.

The time required for any electrical operation, such as turning a current on or off, depends on the *capacitance* and *resistance* of the cable or wire. Resistance and capacitance are usually proportionate to the distance from transmitter to receiver. When a signal is put on a wire at one end, it does not immediately appear at the other end even though electricity travels at the speed of light. The signal increases from zero to the maximum value slowly, as shown in Figure 2-4(b). If the signal at the transmitting end is impressed and taken away in too short a time, a very small or hardly noticeable change will appear at the receiving end. The signal may not be *detectable*.

Remember the word detectable; it is crucial to the telecommunication of information. It is natural to think that there would always be some signal change at the receiver end when a signal is transmitted. However, no signal is detected unless it is greater than a certain *threshold*. Threshold is another important concept in understanding how information is transmitted. Threshold depends on both the sensitivity of the receiving equipment and the magnitude of the noisy fluctuations which, as we have noted, always occurs in any communications channel in our electrically noisy world. If the signal is too weak with respect to the noise, the signal transmitted will not be detected. This

FIGURE 2-4
Distorted and Noisy Signals.

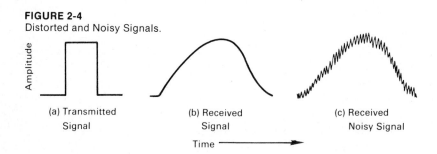

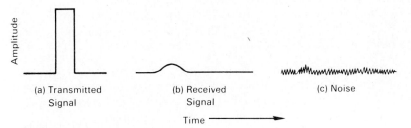

FIGURE 2-5
A Signal Below the Threshold of Detection.

is shown in Figure 2-5. The transmitted and received signals are shown as voltage (a measure of the current in the signal) with respect to time. At the transmitting end, the voltage is turned on and a signal impressed (Figure 2-5(a)). At the receiving end the voltage begins to rise, arriving at some level which is always less than the transmitted level. When the transmitted voltage is switched off, the received voltage slowly drops to zero (Figure 2-5(b)). Now noise is added in the channel. Note that when the transmitted voltage is turned on and off too fast the received signal does not have a chance to rise above the threshold level; it will be undetected (Figure 2-5(c)).

Noise

Noise is always present and unwanted. In your travels through the world of telecommunications you will often hear engineers and salespersons talk about signal-to-noise ratios which are measured in decibels (a tenth of a bel, nee Bell). A signal-to-noise ratio of ten decibels (db) means that the signal power is ten times the noise power. A signal-to-noise ratio of 20db means that the signal power is 100 times the noise power. A ratio of -3db means that the signal power is half the noise power (Those who are mathematically inclined will have surmised that these ratios have something to do with logarithms. The number of decibels is ten times the logarithm to the base 10 of the ratio.)

The Designer's Dilemma

In our study of communications technology you will see that engineers and users are always concerned with improving the cost effectiveness of the communication channel. Many of the major developments in communications technology over the past thirty years, including the satellite and the microprocessor, have focused on how to reduce the cost of sending a message and how to ensure that the information sent is accurately received. One way to reduce the cost of transmission is to increase its speed. We have found, so far, two factors that limit the speed of communication.

1 There is always noise on the communication channel. The received signal must be greater than the spurious fluctuations due to noise; and

2 The received signal is always attenuated—that is, the received signal is always less than the transmitted signal. If the signal is attenuated too much it will not be detected above the noise.

The primary engineering problem in the design of telecommunication systems is how to deal with distortion and noise so that the received signal reproduces the transmitted signal as faithfully as possible and at low cost. This book discusses how engineers are trying to solve this problem.

THE TELEGRAPH AND DIGITAL COMMUNICATIONS

The telegraph is a most interesting instrument. Dating back more than 150 years, it is one of the earliest telecommunication systems. Yet it operates in a thoroughly modern manner—that is, it is a digital system! The Morse Code actually uses three symbols—the dot, the dash, and the time spaces that separate them and enable us to distinguish the dot from the dash. If we were to eliminate the dot and use only the dash and if we were to call the dash a "one" (1) and the time space a "zero" (0), we would have today's digital code which the computer uses as the symbols of its language. Digital communication systems use just two symbols, a one and a zero. The presence of a tone or voltage is represented by a "1" and the absence of a tone or voltage is represented by a "0."

The process of sending information along a wire by means of electrical impulses is, as we have seen, fairly straightforward. Still, engineers must spend a great deal of intellectual energy and investor dollars, francs, marks, rubles, and pounds overcoming spurious noise and distortion to ensure that the information sent is the information received. But how is information sent by "wireless," or radio? This method requires a different kind of signal, one that can travel through the air or, as broadcasters say, over-the-air.

WIRELESS COMMUNICATIONS

We pause now to describe one of the more complex but important concepts in telecommunications—that of transmitting information at a distance without wires, what is commonly known as broadcasting.

The Greeks observed that certain materials had the property of affecting other materials at a distance, through the air, and without any physical connection. The magnetic stone from Magnesia in Asia Minor, for example, produced an effect which could be experienced by bringing another of these lodestones nearby.

In the early days of electrical investigations it was observed that an electrically charged particle—a grain of sand on which an electrical charge had been placed or a small metal ball suspended in air with a charge on it—exerted a force on another charged particle placed near it or in its *field*. The sand particle or metal ball produced an electric field which acted at a distance on other charged particles nearby without any physical connections between them, just as the magnetic stones from Magnesia had done.

Whether at rest or moving, these particles or stones produce electrically charged *fields*. When at rest, the charged sand particle or metal ball produce electrostatic fields, whereas the stones, or magnets as they came to be called, produce magnetostatic fields. If we are able to get the electrostatic field moving or changing (by moving the metal ball or sand particle about or changing the amount of electric charge on these items, for example), the electrostatic field produces a field around it that is similar to the field produced by a magnet. The resulting *electromagnetic field* is both electrical and magnetic.

Michael Faraday, who might very well be called the father of electrical engineering, observed that when an electric current is passed through a wire an electromagnetic field is produced in the space around the wire. This is to be expected; an electric current is essentially a stream of charged particles, or electrons, moving in the wire. The electromagnetic field surrounds the wire just as it surrounds the charged moving sand particle or the charged swinging metal ball. This field theoretically extends to an infinite distance. In reality, however, the further you are from the wire, particle or ball, the weaker the strength of the electromagnetic field and the less its effect.

If another wire is brought into this changing or moving field, a current flows in that wire that is similar to but much weaker than the current in the first wire. This is exactly what happens when a signal is transmitted from one place to another. The signal originates in an antenna or wire. At a distance, another antenna or wire receives and reproduces the signal in a form that is similar but weaker than the signal in the originating wire. To transmit a telegraph signal over-the-air it is necessary to produce a varying signal in the first wire that can generate fields with enough strength to reach a distant wire. The simple dot-dashes of the telegraph will generate only very small fields capable of traveling distances of centimeters or inches, and only if there is a very sensitive device or detector nearby will a transmitted signal be recognized.

Electrical engineers, however, have found that electrical vibrations that move very rapidly can travel over-the-air for long distances. These are carriers which can be used to broadcast messages such as Morse Code. The sine wave, or sinusoidal signal, shown in Figure 2-6(a) has become the universal carrier because it describes many familiar vibrations in nature. The simple harmonic motion of the sine wave can be used to describe water ripples, a slowly swinging pendulum, and a tuning fork sounding pure musical tones. For the engineer the sine wave is a blessing; it is mathematically easy and convenient with which to work.

In order to not confuse the recipient of your message, you must be certain that the signals you send will be easily differentiated from the underlying carrier necessary for transmitting over-the-air. For this reason, you need a carrier that not only vibrates rapidly (is of a high frequency) but one that is very regular and predictable. The sine wave is just such a carrier. In Figure 2-6(b) the Morse Code is easily recognizable even when superimposed on the carrier.

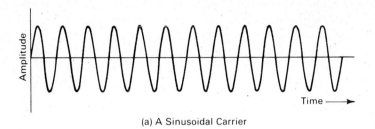
(a) A Sinusoidal Carrier

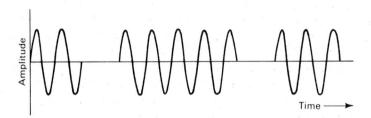
(b) Morse Code on the Carrier

FIGURE 2-6
Morse Code Modulating the Carrier (On-Off Keying, OOK).

From here on much of this chapter will discuss how these carriers are used for the transmission of information signals.

MODULATION

Let us see how our modern telegraph signal can be sent over-the-air. First we have the message in digital form, or, if you remember, in our dash-space format as shown in Figure 2-7(a). Following the digital terminology of binary notation we have labelled the dash a "1" and the space a "0." We know that the signal will be distorted even before it gets into the transmission system for it must pass through wires and devices prior to joining the carrier for its long-distance transmission. So in Figure 2-7(b) we see the familiar shape of a signal that has been affected by certain losses. Figure 2-7(c) shows what happens when our signal is combined with or superimposed on the underlying carrier; this is called *modulation*, the process of impressing the signal on the carrier. The carrier is mediated by turning it on and off in accordance with the signal or information we are sending by our modern Morse Code. In the "trade," this modulation technique is called *On-Off Keying (OOK)*. Because we are changing the amplitude, or power, of the carrier from a value which represents a "1" to one that represents a "0," OOK becomes a form of *amplitude modulation*.

Now to make this example more realistic, let us assume that there is some noise on the transmission line. Figure 2-7(d) shows what happens when the now noisy signal is *detected*, or separated, from the carrier. We can see the highs and lows of our ones and zeros (the dashes and spaces in our Morse Code message), but they are cluttered with the noise that rides on top of our signal. The noise can be separated from the signal by *filtering* (separating the higher frequencies from the lower frequencies); our message is shown in Figure 2-7(e).

The equipment used to display our message, the *decoder*, responds only to ones and zeros and what it finally delivers to us is shown in Figure 2-7(f). Note what has happened. The presence of noise during transmissions has led to errors; some of the zeros are received as ones and some of the ones as zeros!

FIGURE 2-7
Signal Transmission in the Presence of Noise.

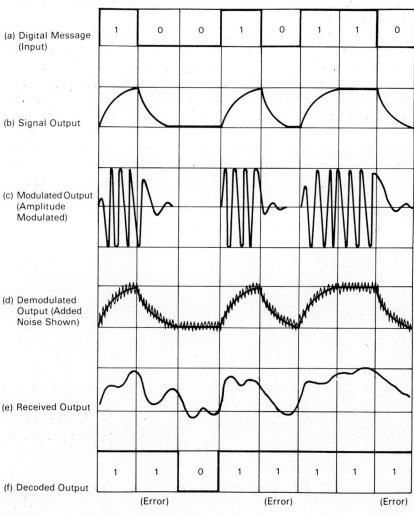

Another method for sending signals is to vary the frequency of the carrier as shown in Figure 2-8. The lower frequency (slower vibrations) would represent a zero while the higher frequency (faster vibrations), a one. This kind of modulation is referred to as *frequency shift keying (FSK)*. It is a form of *frequency modulation* which is the basis of high-fidelity music systems. Frequency modulated (FM) signals are less sensitive to noise than are amplitude modulated (AM) signals. Because the likelihood of an error is less, FM signals are often used for data communications. The ability to send data, or digital signals, over the telephone, which is not a digital system, is made possible through the use of FSK.

There is a third technique for modulating an underlying carrier in order to broadcast information and, while you will be hearing more about it later when we begin to deal with data communication systems, it is well to fill out our bag of modulating tricks now. This technique involves changing the *phase* of the carrier in accordance with the signal rather than the carrier's amplitude or frequency. Put simply, this means altering the timing of the carrier when you start the signal moving. Note that in Figure 2-8 the one and zero signals begin at the same point in the sinusoidal wave (carrier)—at the start of the upswing of the curve. Now examine Figure 2-9. Here we begin the zero at the bottom of the curve and the one at the top of the curve. The signal is shifted to the right for the zero and to the left for the one. This is *phase shift keying (PSK)*. PSK is even more immune to noise than FKS but requires more complex equipment and is often more expensive than FSK or OOK.

Rather than begin at the usual beginning with the technology of the telephone which would have led us to the transmission of speech, we began instead at the very beginning of telecommunications history, with the telegraph. This led us to the transmission of digital signals rather than voice signals and to the realization that the transmission of dots and dashes, or ones and zeros, is not at all difficult. In subsequent chapters you will learn that there are a great many benefits to transmitting digital representations of continuous signals, such as voice. These benefits include less concern about noise when using FSK and PSK, and the ease with which you can regenerate rather than amplify digital signals when they are sent over very long distances.

FIGURE 2-8
Frequency Shift Keying (FSK).

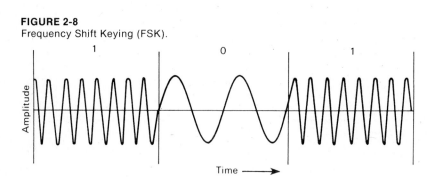

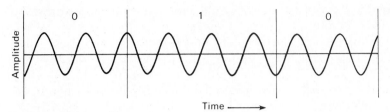

FIGURE 2-9
Phase Shift Keying (PSK).

Two questions come to mind immediately. Since digital transmission is more likely to give us reliable information transfer, why didn't we build on Morse's telegraph and "go digital" for our telephone? And, if signals represent *information*, what exactly is information?

The answer to the first question is easy. The neat and tricky devices that now make digital communications so simple were not available a hundred years ago; we needed the semiconductor chip to make this possible. The second question requires a somewhat more detailed answer.

WHAT IS INFORMATION?

Suppose we want to transmit this page to someone in another city. There are several ways to do so. We could make a long-distance call and read the page over the telephone. We could rent time in a local TV studio, arrange to place the page before a camera, and transmit a television picture of it. We could code the page in Morse and telegraph it to the distant city.

This page has about 2000 characters (letters, numbers, spaces, and punctuation marks) on it. These 2000 characters contain its "information." If we were to send the page by Morse Code we would need two symbols with which to describe each character or a total of 4000 symbols. (Remember, Morse Code uses two symbols, a dot and a dash.)

The symbols we are going to use to describe the characters on this page, the ones and zeros we assigned to the Morse Code, have been given the name "bit," which is short for binary digit. The term binary digit comes from the fact that just two digits are used.

How many bits—zeros and ones—do we need to send this page via a modern telegraph or Telex system, over the telephone, or over a television channel? This is an important question because each of these methods of sending this page can only accommodate a certain number of bits in a given period of time. Remember how the transmitted dot or dash looked when it was received? And if the transmission was too fast how the signal might never rise above a certain threshold and thus would not be detected?

A modern telegraph system or Telex transmits at about a rate of sixty words per minute and uses a five-bit code for each character. If we assume that there are about thirty bits per word, then we would have to transmit approximately 200,000 bits in one minute.

The telephone line, at best, can transmit about 9600 bits per second. So, if we were to read the page over the telephone in one minute, we would need 576,000 bits.

A television frame consists of 525 lines with about 500 picture elements per line. (A picture element can be thought of as the photographer's way of measuring different shades of gray or color and can be coded or translated into bits.) The video frame will be transmitted every 1/30 of a second or 1800 times in one minute, each time transmitting 1,575,000 bits. In the one minute that it takes to read the page, 2,835,000,000 bits will be transmitted!

Each method of transmission uses a different number of bits to send the same message, the same amount of information, in the same period of time. This raises several very interesting questions:

Is there a maximum number of bits a communications channel can send, in other words a maximum channel capacity?

How do you measure information?

What is information?

Most of us think of information as what we obtain by asking for directions, talking to friends, reading newspapers, listening to the radio, watching television, and dialing "information." When using the telecommunication system, it's *the type* of information that is important to you. But when engineers design systems they must be concerned with *the quantity* of information. This is the critical distinction between how social scientists deal with information in their study of human communications and how the engineer must define information in order to design telecommunication systems for human communications. It is the beauty of the mathematical theory of communications that a bridge is created between these two definitions.

If the flashlight doesn't blink, the semaphore doesn't move, the television picture doesn't change, or the voice tones over the telephone do not vary, then no information is conveyed. When we blink the light, wave the semaphore, move the picture, and speak different sounds into the telephone, information is conveyed.

If every month we were told that the prime interest rate has gone up again, pretty soon we would quit asking about it. But if we were told interest rates went down last month, that's new information. Information is conveyed only when there is doubt and uncertainty present, and the information can remove the doubt and resolve the uncertainty. Information that resolves greater uncertainties or allows us to make better choices among many alternatives is valued more highly than information that does not.

In 1948, Claude Shannon of the Bell Laboratories showed that the mathematical theory of information did, indeed, agree with human experience and common sense. He pointed out that:

1 For a signal to carry information, the signal must be changing; and

2 In order to convey information, the signal must resolve uncertainty; the greater the uncertainty the greater the information contained in its resolution.

HOW DO YOU MEASURE INFORMATION?

Have you ever inadvertently played a 33 1/3 RPM record at 78 RPM? Consider what would happen if you allowed the record to squawk its way to the end. First of all, it would complete its play 2 1/3 times faster than if played at its required speed and you would receive all of its information 2 1/3 times faster. Second, the music tones and frequency of sounds would be higher pitched. While the normal range of frequencies of your record might be between 20 and 20,000 cycles per second, running faster the record would produce frequencies perhaps up to about 47,000 cycles per second or 2 1/3 times higher.

It would seem, then, that if you wanted to send a given quantity of information twice as fast you would simply have to double the frequency range or *bandwidth* of your channel. In other words, a definite amount of time and bandwidth is required in order to transmit a given quantity of information. Information, then, can be measured by time and bandwidth.

Remember the term *bandwidth*, you will be hearing a great deal about it and about the fact that *bandwidth costs money*. When you installed a central switch in your network (see Figure 2-2), you may not have saved money by substituting a single complex switch for many smaller and less complex switches, but you probably eliminated many miles of cables and wires. Copper transmission lines are expensive providers of bandwidth, so you have conserved bandwidth.

Bandwidth also could be used more efficiently by sending dots and dashes more rapidly along the telegraph line, enabling you to send many more messages per hour. But this would be the same as increasing the speed of your record; you would be increasing the frequency range and for a higher frequency you would need more expensive bandwidth. In our exercise on transmitting a single page we noted that different transmission systems could only send so many bits per minute. They were bandwidth-limited. And don't forget the other limitations on the speed of sending bits, or dots and dashes: the distortion due to the characteristics of the transmission system and our ever-present noise. Let's not forget, too, that in order to detect the signal from the noise, we must have a suitable signal-to-noise ratio.

IS THERE A MAXIMUM CHANNEL CAPACITY?

Not long after the first telegraph lines were constructed, users began to search for ways to send more than one message at a time. The line was rather expensive even in those days and, if the only way to send more information was to string another line, the cost to send messages would soon rise far beyond what people were willing or could be expected to pay. So early in the history of electrical communications the capacity of a telecommunications channel to send information was of concern to the providers and users of telegraph services. How to make more efficient use of this capacity became a major engineering challenge and, indeed, remains so today.

The crowning achievement in the linking of the mathematical theory of communications to the social scientists' definition of information is the ability to quantify the parameters that enable us to determine the maximum information capacity of a communication channel. Shannon's principal contribution to communication theory was the development of a means to determine how much information can be transmitted in a period of time in the presence of noise. This measure, which is so important to our efficient use of telecommunication systems, is determined by four measurable parameters: time, bandwidth, signal power or strength, and noise power or strength.

Throughout the remainder of this chapter we will discuss how to make ever more efficient use of valuable communications bandwidth. The mathematical theories of communications are powerful and practical tools for both designers and users of modern telecommunications. They define the limits of our search for the most efficient use of communications bandwidth, our primary resource for information transfer.

Our first stop in this search for communications efficiency will take us away from the digital world for a while, for despite the rapid increase in data communications, we still live in an analog world. We next explore how to make the best use of analog communications channels by way of the many ingenious and important techniques for sending multiple signals on the same channel. We shall return to the digital world at the conclusion of this chapter to discuss why digital communications is the wave of the future.

TRANSMITTING SPEECH

Newspapers and magazines are full of news about the explosive growth of computers and data services and the emerging communications network for shopping, banking, selling, learning, and working. This is the network marketplace we mentioned briefly in the first chapter and which we shall visit again in Chapter 13.

Daily living tasks such as these are best done on a digital telecommunications network and it would seem that we are on the brink, if not in, the digital era. But this is not the case. Indeed, even by the end of the century, the bulk of the traffic on our networks will be taken up by human communications—people talking to people. We are a very talkative society and as our population increases we, and especially our children, will transmit voices over telephone lines.

When we speak or sing, our vocal tract produces pressures in the air around us which are recognized as sound waves. When we speak into the telephone, the air pressure our voice produces impinges on the microphone in the mouthpiece of the telephone handset which, in turn, converts these pressure variations into electrical or current variations, or vibrations. The receiver handset in the earpiece converts electrical variations coming back to us into sound waves which impinge on our eardrum. The electrical signal is the physical *analog* of the sound waves produced by the human speaker and heard by the human listener. The signal being transmitted is called an analog (as opposed to digital) signal because it is analogous to the voice sounds. The telephone is an analog communications system; the telegraph is a digital system.

We live in an analog world. Your telephone, radio, television, high- (or low-) fidelity music systems, and all of the new video marvels are analog systems. Although our old and by now friendly telegraph has been operating for more than 150 years as a digital system, the sharp break in our analog world has come about because of the computer. We shall have much more to say about digital systems later, but for now, we must understand how our present world—the world of analog telecommunications—works.

The sound waves our vocal tract produces are quite complex. While a tuning fork produces a sound wave at only one rate of vibration and a chord played on the piano or guitar may have several rates of vibrations. The human voice is really composed of many vibrations of different rates.

Rather than continue to refer to vibrations, let's get used to the technical term, frequency. Frequency is defined, simply, as the number of vibrations or cycles per unit of time. Thus, if a tuning fork is vibrating at 440 vibrations a second, it has a frequency of 440 cycles per second.

We cannot say that a voice has a frequency of, say sixty cycles per second (or 60 cps) because it is made up of so many frequencies. How, then, can you talk about the "frequency" of someone's voice? To transmit the voice, we need to know something about its "frequency" so that an efficient transmitting system can be selected or designed. We do so by means of that very important parameter previously mentioned, *bandwidth.*

BANDWIDTH

We can "compose" a complicated voice signal by adding up a number of standard and well-defined signals—our familiar sine wave, for example. Mathematicians, and especially Joseph Fourier, long ago showed that by combining a number of sine waves of different frequencies, amplitudes, and phases (a change in phase, as we have seen, corresponds to shifting the signal to the right or left on the time scale), it is possible to construct waves of any shape and size. The number of sinusoidal signals making up the sum is infinite, but because we have difficulty dealing with infinity in our lives, we do not count those sinusoidal components that are very small in amplitude. What is truly wonderful about Fourier's work is that he demonstrated that there is only one way to compose any desired signal—that is, only one selection of frequencies, amplitudes, and phases determine a complex signal.

The range of frequencies of the component sinusoidal signals used to compose the voice signal—in other words, the difference between the highest and lowest frequency used—is called the *bandwidth* of the signal. Figure 2-10 shows how this looks.

Bandwidth is an extremely important characteristic because the cost of a signal's transmission depends fundamentally on its bandwidth. Generally speaking, higher bandwidth signals cost more to transmit. Because bandwidth is the difference between the cycles per second of the highest frequency and that of the lowest frequency making up the signal, it is also measured in cycles per second. In honor of Heinrich Hertz, who produced the electromagnetic waves about which Clerk Maxwell theorized, we now call a cycle per second a hertz. Today's frequencies are getting higher and higher and bandwidths are getting broader so you will be reading and talking about kilohertz (KHz) or 1000 cycles per second, megahertz (MHz) or 1,000,000 cycles per second, and gigahertz (GHz) or 1,000,000,000 cycles per second.

Here are bandwidths for some common signal transmission systems we have been talking about:

Telegraph	40 hertz
Telephone	4000 hertz
Hi-fi music	20,000 hertz
Color television	6,000,000 hertz

Different communication channels transmit signals best at different frequencies; they have different bandwidth capacities. Consider the simplest

FIGURE 2-10
Bandwidth Is the Difference between the Highest Frequency Component and the Lowest Component (BW = $f_{max} - f_{min}$).

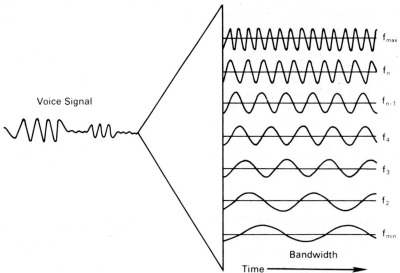

channel of all, the air space between two people speaking to one another. They can understand each other if they hear one another at the normal frequency of speech. The air does a fine job of transmitting the sound pressures formed when we speak. We can extend the distance between the parties speaking by using loud-speaker systems which can be either mechanical (a megaphone) or electrical (an amplification system). But if we want to transmit speech over very long distances a channel or transmission mode better suited for the long-distance transmission of speech or any other signals containing information is required.

In our previous discussion of *broadcasting* we noted how signals can be sent over long distances. To send voice signals with a bandwidth of between 40 hertz (Hz) and 4000 hertz (Hz) would need very large devices (antennae) and an enormous amount of power. We have to, somehow, "alter the air" if you will, and make it more amenable for the transmission of speech. We referred to this process of alteration as *modulation*. It is the process used to transform a signal into a form which is well suited to the channel for its distribution. *Demodulation* is the process that recovers the original signal from the transformed one for delivery to the receiver.

In Figure 2-7 we showed the process by which a digital message is transformed (modulated) so that it will fit a channel for transmission. We shall now discuss modulation in terms of analog signals and describe the world in which we live today with its telephones, radios, and television sets.

AMPLITUDE MODULATION

Let us work with Figure 2-11. In this figure, (a) is said to modulate the higher frequency underlying carrier (b) and the resultant modulated signal is shown in (c). This is a curious way of putting it; it would seem that the signal is really modulated by the carrier, hence we say the modulated signal. In effect, it makes little difference because what is transmitted is a combined signal where the original signal, our information, can be mapped by connecting the peaks of the modulated carrier (both above and below). The space in between is known as the amplitude envelope of the modulated carrier as shown in (c).

It wasn't very long before communications engineers recognized that this was not a very efficient way of using bandwidth. They came up with the ability to transmit two signals on one carrier by using both the upper and lower portions of the envelopes (the sidebands) for two different messages. Because transmitting via sidebands is becoming increasingly important, especially for mobile radio communications and paging, we shall stop for a moment to find out just what these sidebands are.

We defined frequency as measured by the number of cycles per second. Frequency is defined by time; hence there should be a relationship between time and frequency and, indeed, there is. If you were to show frequency as related to time, the vibrations would move, as in Figure 2-12(a). In this figure we have a signal with a single rate of vibration that is stable and uniform; it is a pure tone from a tuning fork. It has one frequency value which may be seen on

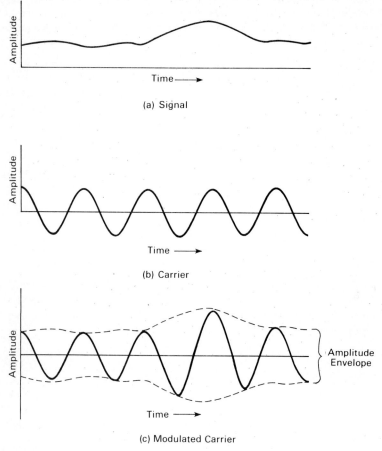

FIGURE 2-11
Amplitude Modulation (in the Time Domain).

Figure 2-12(b) as a single point on the frequency line or axis. Now, if the signal is a more complicated one, such as a human voice, as in Figure 2-12(c), it would be made up of many frequencies—that is, it would have a large bandwidth as in Figure 2-12(d).

What would the time side of Figure 2-12 look like if we were to substitute for our voice or pure tone, a pulse or the Morse Code dash (see Figure 2-12(e))? On the frequency side of Figure 2-12(f) is a whole range of frequencies of many values, from the lowest or zero cycles per second at the start of the pulse to, perhaps, many thousand cycles per second at the end of the pulse. The bandwidth of the pulse is very much higher than the bandwidth of our voice signal.

We can generalize this observation by noting that the faster a signal changes the more bandwidth is necessary to transmit it. Thus, to send many short

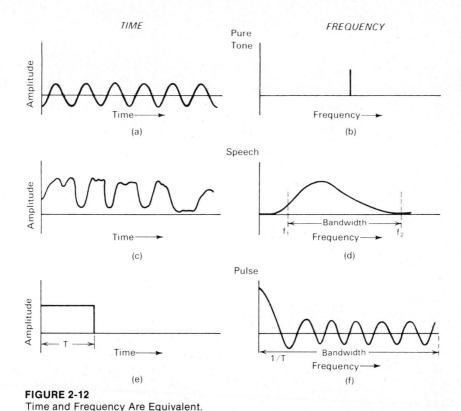

FIGURE 2-12
Time and Frequency Are Equivalent.

pulses, a larger bandwidth than even the best of the high fidelity music systems now on the market would be required.

Bandwidth is a central issue in the design of telecommunication networks. If we are to make the most cost-effective use of bandwidth, we must send more than one signal in a particular bandwidth. But by doing so, we produce a great many signal variations or frequencies and that increases bandwidth. Indeed, the situation is likely to be made even more difficult in the digital world for, as we have just seen, pulses take up bandwidth. You have just been exposed to the central problem of communications engineering.

Now let us go back to amplitude modulation (AM). What does our AM signal look like on the frequency side? Here is our signal in Figure 2-13(a) with a bandwidth of B hertz. Here is our underlying carrier in Figure 2-13(b), at a pure frequency of f_c cycles per second much higher than B hertz. And here is our modulated carrier in Figure 2-13(c) with the information signal bandwidth both above and below the carrier. Note that they form bandwidths which we call lower and upper sidebands.

Engineers now use these sidebands to transmit two different signals, thereby making more efficient use of valuable bandwidth. In particular many military radio communications are carried on only one of the two sidebands, leaving

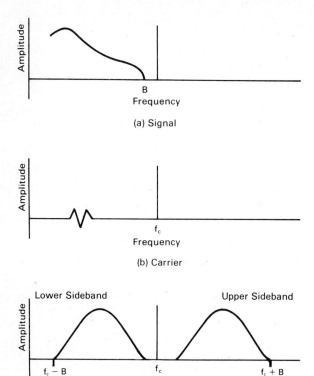

FIGURE 2-13
Amplitude Modulation (in the Frequency Domain).

the other sideband for different signals, hence the term "single sideband radio transmission." This is a simple but nevertheless important way to make better use of valuable spectrum. We shall talk more about how to make efficient use of bandwidth elsewhere in this book.

FREQUENCY MODULATION

In frequency modulation, or FM, the carrier *frequency* is varied with changes in the information signal to be transmitted. To illustrate, let us use a rather odd but important signal engineers call a "sawtooth" wave. It is the triangular signal shown in Figure 2-14(a) used in television transmission to sychronize the transmitted information with the receiving information on your set. The amplitude of this sawtooth signal modulates the frequency of the carrier and changes it as shown in Figure 2-14(b).

From our previous discussion about how much bandwidth a rapidly changing signal uses, we can see that the sawtooth signal is a devourer of bandwidth; a

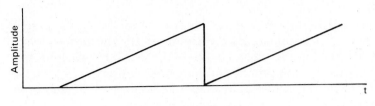
(a) "Sawtooth" Modulating Signal

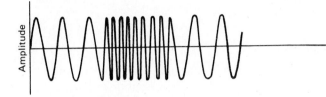
(b) Frequency Modulated Carrier (FM) by the "Sawtooth" Signal

FIGURE 2-14
Frequency Modulation.

signal which rises in a short period of time will generate very high frequencies. To conserve bandwidth we don't usually send all of the frequencies, just enough to clearly define the information. In other words, how much bandwidth a frequency modulated signal uses depends on how much you choose to vary the frequency of the carrier. In FM radio, stations have been allocated 200,000 hertz around their assigned frequency by the Federal Communications Commission.

How do you select between AM and FM? What are the benefits and drawbacks of each? Here are some rules of thumb:

• In FM, you can increase the bandwidth and improve the signal-to-noise ratio without increasing signal power.

• In AM, you can only increase the signal-to-noise ratio by sending a more powerful signal.

• FM receivers are somewhat more complicated than AM receivers and, consequently, more costly.

• FM transmitters are usually more difficult to build than most AM transmitters.

So take your pick of benefits. With the continuing reduction in the cost of electronic components, the difference in price between FM and AM receivers has all but disappeared, but the cost of bandwidth is still high. However, if you can pack a great many signals into this bandwidth or channel, the transmission, costs per signal could be low. This might tend to favor AM, but at a cost of more power. Such are the problems of design!

MULTIPLEXING

Several times we have mentioned the possibility of sending more than one signal on a channel at the same time. This can be done by shifting the frequency of the signal thereby giving each signal a different place in the channel and thus sharing the spectrum. This process is called *multiplexing*.

If it were not possible to multiplex, our telephone systems would require an enormous number of wires and at today's copper prices, phones would be extremely expensive. Even with extensive switching systems we would still require a large number of wires, one for each potential simultaneous conversation between two points.

FREQUENCY DIVISION MULTIPLEXING

By multiplexing several signals in the same channel by means of frequency separation or *frequency multiplexing,* we can conserve copper and bandwidth. Figure 2-15 shows how one type of telephone system multiplexes twelve telephone channels. Each signal requires 4000Hz of bandwidth. Using a carrier in the range of 60KHz to 108KHz or a bandwidth of 48KHz the system assigns 4KHz of space to each of the twelve channels, thereby filling up the 48KHz single cable with what would normally have required twelve cables. In "real life" there are frequency spaces or separation bandwidth called guard bands between each of the twelve channels so that the system would probably require somewhat more than 48KHz to carry twelve simultaneous conversations.

Multiplexing also saves money by allowing for fewer numbers of wires or cables between major population centers. Indeed, because variations in popu-

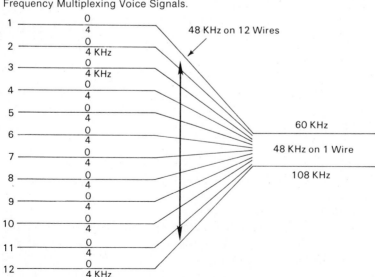

FIGURE 2-15
Frequency Multiplexing Voice Signals.

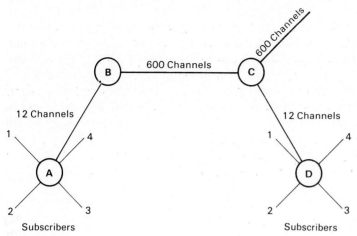
FIGURE 2-16
Frequency Multiplexing Distributes Telephone Signals More Efficiently.

lation density make telephone traffic density very different in different parts of the country, it is uneconomical to provide a facility for, say, 600 channels when only ten to twelve channels are likely to be in use at any one time. But a single 600-channel system costs less than fifty 12-channel systems due to savings in the number of transmission lines and amplifiers. Thus, there is a need for a number of transmission systems, each optimized for a particular channel capacity. This is made possible by multiplexing and is illustrated in Figure 2-16. The more signals you send the more frequencies you need and this increases the bandwidth necessary for transmission.

TIME DIVISION MULTIPLEXING

As we have learned, information transmission requires the allocation of both time and frequency—that is, the quantity of information transmitted depends on both bandwidth and time. Frequency division multiplexing divides up bandwidth, but you can also divide up time. How these approaches differ from one another is illustrated in Figure 2-17. Note that frequency multiplexing takes up a certain bandwidth during the entire time of transmission. *Time division multiplexing* (TDM), on the other hand, uses the time of transmission sporadically, taking up small portions of the time with a signal or many signals. The value of TDM is that signals can be mixed in a time period by separating them in time, as long as you can keep track of which signal is being sent when in order to reconstruct your messages. Today, time division multiplexing is, in its many forms, the most promising means for low-cost transmission of information and for the cost-effective use of bandwidth.

The semiconductor chip with its ability to perform many electronic functions extremely rapidly because of its size (and many other factors which we shall

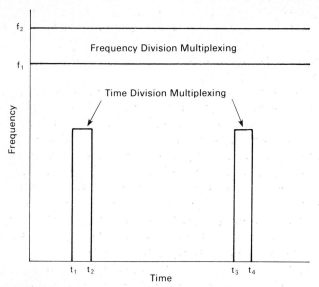

FIGURE 2-17
Time vs. Frequency Multiplexing.

discuss in the next chapter) and its low cost considering all it can do, is primarily responsible for the recent development in TDM. To be able to send analog signals via TDM requires that we select portions of the signal to be sent without in any way destroying the meaning of the information being communicated. This requires a process known as *sampling*.

SAMPLING

We have seen how a continuous signal, such as a voice signal, can be used to modulate an underlying carrier. But we shall see that some good reasons exist for using a different sort of underlying carrier—a string of pulses, for example.

Figure 2-18(a) shows a continuous signal with the information we want to transmit. This time, however, we wish to modulate a string of pulses as shown in Figure 2-18(b). Now look at Figure 2-18(c). We have modulated the string of pulses (pulse train) in accordance with the information signal Figure 2-18(a). The signal is pretty well represented in the modulated carrier. It is as if we were connecting points on a data curve; essentially, we have passed over the spaces between the points on the assumption that the missing spaces do not contain any important data points. In most cases we would be entirely correct and would not seriously misrepresent the meaning of data.

Look at Figure 2-18(c) and note that there are empty spaces in the modulated pulse train. What if we could fill these spaces with other pulses that represent other analog signals? Wouldn't that allow us to send many signals on the same

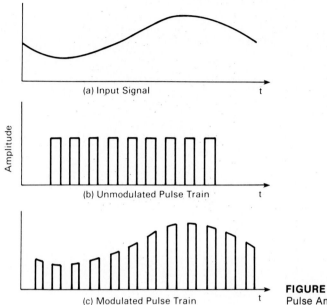

FIGURE 2-18
Pulse Amplitude Modulation.

channel? To do so would require that we take our continuous analog signal and break it up into a string of pulses that still represents the information we wish to transmit.

This is essentially what we do when data is gathered about some continuous natural phenomenon or some aspect of human behavior that is continuous and these points are connected on a graph in the hope that they will, indeed, represent what is really taking place. We sample reality and then hope that the number of samples can be a sufficiently accurate picture of reality. To sample signals for transmission and still maintain the accuracy of the original signal at the receiving end requires that we do three important things:

1 Determine the sampling rate—that is, the number of pulses per second needed in order to represent the signal accurately,

2 Have the equipment that can perform this sampling rapidly and accurately, and

3 Have the facilities to recapture the original signal from the pulses accurately and, of course, in the analog form we need in order to interpret the message.

PULSE AMPLITUDE MODULATION

We propose to transmit a signal, an analog representation of, say, a singing voice. This information is shown in (a) of Figure 2-19. If we sample at a very low rate, as shown in (b) it is clear that the original signal is not likely to be

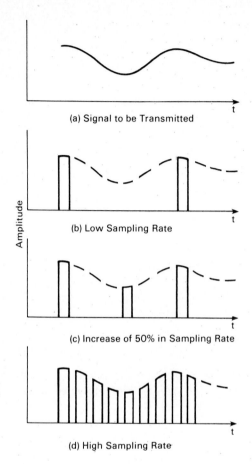

FIGURE 2-19
Sampling for Pulse Amplitude Modulation.

reproduced, or it will be reproduced in some very imaginative but inaccurate ways. Remember, the signal in Figure 2-19(a) is not a straight line which can be determined or defined by just two points. Now let us increase the *sampling rate* by 50% or by sampling another point on the curve as shown in Figure 2-19(c). It is a bit more likely that the original signal will be recognizable, but there is still a great deal of room for imagination. Finally, we increase the sampling rate by a factor of a hundred or so as shown in Figure 2-19(d). The original signal is very clearly and quite accurately reproduced. If we sample at a higher rate, say increase the sampling rate by a thousand times, we shall not add much to the quality of the reproduced signal. So there must be some optimum or best rate of sampling that will give us the accuracy we desire.

Sampling at a very high rate adds little to the intelligibility of the signal. Sampling at a very low rate risks missing some of the variations in the signal, in essence, missing the information being transmitted. The key requirement is that the sampling rate pulse rate, or frequency rate of sampling, be at least twice the highest frequency in the information signal itself. This is the famous Sampling Theorem of Hartley and Shannon. Thus, for a voice signal with a bandwidth of 4000Hz the sampling rate should be at least 8000Hz. For a color television signal of 6MHz, we must sample at a rate of at least 12MHz.

With the ability to create a pulse train that accurately represents the information being sent, we can modulate a carrier which will transmit the signal. We have described yet another form of amplitude modulation, but in the case of *pulse amplitude modulation* we are modulating a train of pulses in accordance with the signal amplitude rather than a continuous sinusoidal underlying carrier.

We can send information in a way that does not use up all of the time allotted on a channel to a single message. Indeed, as you can see in Figure 2-19(d) there are many unused time spaces. This is an application of *time division multiplexing* which was defined previously.

Suppose we have three different messages to send. The signals representing these messages are shown as A, B, and C in Figure 2-20(a). We sample each of these signals at different times as shown in the figure and then put all of the resulting pulse trains together in a string as shown in Figure 2-20(b). If each of these signals has a different bandwidth, we shall be sampling at a rate determined by the signal with the largest bandwidth.

If we recover or demodulate these pulses in the proper sequence, we can reproduce the original signals, A, B, and C. At the receiving end of the

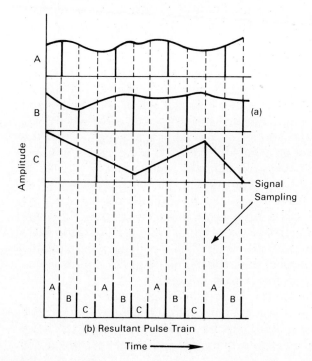

FIGURE 2-20
Pulse Amplitude Modulation.

transmission we shall need equipment that will operate in synchronization with the equipment that sampled the original signals.

Pulse amplitude modulation (PAM) systems have become very attractive for sending multiple voice signals along a single transmission channel. PAM systems ought not to be confused with digital systems; they are not completely digital since the amplitudes of the pulses transmitted vary continuously with the original analog signal variations. The job of recapturing the several streams of pulses carrying information is made difficult by the need to precisely synchronize the timing at the receiving end with the timing at the sending end. One way to ease this process is to digitize the PAM signals before transmission and assign a code to them. At the receiving end you then search for coded signals that come from the same message. This process offers yet another mode of modulation, *pulse code modulation (PCM)*.

PULSE CODE MODULATION

In order to digitize our PAM signals, we must first *quantize* them. To quantize means to break the amplitudes of the PAM signals up into a prescribed number of discrete levels, clearly fewer than the infinite number of levels a continuous signal has.

It is not necessary to transmit all possible signal amplitudes as in a PAM system for two reasons. Because of the noise usually introduced during transmission and at reception, the equipment at the receiving end will not be able to distinguish fine variations in signal amplitude anyway. In addition, our ears and eyes are limited with regard to the fine gradations of signal they can distinguish.

Once values have been assigned to each of the pulse amplitudes, they can be coded digitally. Coding can be quite arbitrary as long as both sender and receiver know the codes. This allows for tremendous flexibility in filling up or loading a channel. The ability to code the signals once they are quantized allows for very efficient use of transmission bandwidth, one reason why PCM is becoming increasingly popular among telecommunications engineers for the transmission of both analog and digital information. Coding signals for transmission is the best known way of overcoming the noise problem.

Remember our telegraph? For that instrument, we made a dash a one and the absence of a dot a zero. Thus, all a receiver has to do when listening for the one or the zero is to recognize the presence or absence of a pulse. The shape of the pulse is not important, only its presence or absence. Nor is the exact amplitude important. By transmitting binary pulses of high enough amplitude we can ensure correct detection of the pulse in the presence of noise with as low an error rate (probability of mistakes) as desired.

Pulse code modulation has another important advantage over other forms of modulation. As a signal is transmitted through a channel whether it be over-the-air or through a wire, the signal becomes weaker while the noise tends to increase. Thus, the signal-to-noise ratio declines the longer the transmission line. Amplitude and frequency modulated systems both require amplifiers or repeaters that will recognize the amplitude or frequency of the incoming signal and reproduce it as accurately as possible, thereby boosting its power for the run to the next repeater.

With PCM transmission it is possible to place very simple devices known as regenerators at intervals along the line spaced closely enough to ensure that the signal-to-noise ratio is fairly high at each interval. The regenerator must only determine if a pulse is present and then regenerate a perfect pulse, or if there is no pulse present, do nothing. The signal-to-noise ratio need not change between transmission and reception!

In the chapters that follow, we shall apply these fundamental concepts to the understanding of telecommunication systems. Our task will be not only to understand how they work but to use this understanding to find out just what they can do for us now and what they are likely to do for us in the future. This chapter has dealt with the fundamentals on which the future is being built. We have tried to provide the "language of the trade" and we hope that you will be able to use this language to create a better trade.

ADDITIONAL READINGS

If there were a rich selection of readings there would be no reason for this book. We have tried to select sources that will enable you to probe further into this topic without being overwhelmed by the enormous technical literature of telecommunications engineering. The following suggestions will enable you to dip your toe as deeply as you may wish into the fundamental concepts required to understand modern telecommunications.

These sources probably contain more than you want to know; we suggest, therefore, that you refer to the indices and cull out the concepts we touched upon and about which you want to know more.

Martin, James, *Future Developments in Telecommunications*, 2nd ed. (Englewood Cliffs, NJ: Prentice-Hall, 1977). This is a rich text that covers a very wide range of technologies. It is an excellent reference.

Pierce, John R., *An Introduction to Information Theory: Symbols, Signals, and Noise*, 2nd rev. ed. (San Francisco: Peter Smith, 1983). A well-written and expert review of significant issues in information theory.

Shannon, Claude and Warren Weaver, *The Mathematical Theory of Communications* (Champaign-Urbana, IL: University of Illinois Press, 1949). The "bible" for information scholars.

Singleton, Loy A., *Telecommunications in the Information Age: A Non-Technical Primer on the New Technologies* (Cambridge, MA: Ballinger Publishers, 1983).

Kellejian, Robert, *Applied Electronic Communications: Circuits, Systems, Transmission* (Chicago: Science Research Associates, 1980). For those who wish to get inside the circuits.

Meadows, Charles T., and Albert S. Tedesco, *Telecommunications for Management* (New York: McGraw-Hill Book Co., 1985).

PART **TWO**

THE TOOLS OF THE COMMUNICATIONS REVOLUTION

Part 2 is devoted to the tools of the communications revolution. These are new tools, and they have caused rapid changes in the technology of telecommunications. Without the semiconductor it is unlikely that we would have moved as swiftly as we have in the application of computing to telecommunications. The massive computers of the 1950s and '60s would probably have remained exclusively in the world of computing rather than migrating as microprocessors to the world of telecommunications.

Without the satellite, the boundaries of telecommunications might not have spread so widely. The satellite has freed us from the constraints of distance and has made communications networks simpler and, consequently, cheaper.

Semiconductors and satellites, inner space and outer space if you will, have created a new architecture for communications, an architecture made possible by innovative devices that provide entree to the network architecture, or terminal.

Our task in Part 2 is to examine the three tools of the communications revolution: the multitalented semiconductor chip, the communications satellite, and the terminals and network architectures made possible by these innovative tools.

CHAPTER 3

THE MULTITALENTED SEMICONDUCTOR CHIP

Next time you are in Washington D.C., be sure to visit the quaint building on the Mall, part of the Smithsonian Museum complex, with this simple sign on its lawn, "1876." Inside you will find a wonderful taste of a technological age long past—the world of the majestic Baldwin locomotive, the Westinghouse turboelectric generator, and the precision machine tools made by German immigrants who settled in Philadelphia and placed their imprint on the "high tech" world that was celebrated at the 1876 Centennial Exposition in Philadelphia. These are the tools of a period historians now call our first industrial revolution.

Now look inside your digital watch, hand calculator, personal computer, or telephone. What do you see? Nothing to compare to the shining machines of 1876. Only tiny globs of plastic, sometimes terribly misshapen, sometimes neat and square, but hardly worthy of display in the Smithsonian, "1986."

These are the tools of the information revolution, the multitalented semiconductor chips that make possible the ⅛-inch wafer that replaces to thirty-ton computer. In this chapter we examine these most important building blocks of modern telecommunications. We discover that not only have the computing and communicating worlds merged, but that communications engineering has returned to the very origins of electricity, indeed to the coaxing of electricity from vials of chemicals as in the early nineteenth century. The design and manufacture of semiconductors are as much chemical as they are electrical.

CAT'S WHISKERS AND SEMICONDUCTORS

The Greeks recognized the material, if not the chemical, nature of electrical phenomena. As noted in Chapter 2, it was the lodestone from Magnesia in Asia Minor that gave us our first indication that some materials had the unusual properties we now find so useful for telecommunications. When Benjamin Franklin and the electricians of the early eighteenth century undertook their often dangerous electrical demonstrations, the chemical nature of electricity became well known. Indeed, their spectacular sparks were produced by electric charges captured in the Leydon Jar, a simple canning jar half covered with tin foil, inside and out. The charge itself came from the same source the Greeks used—the rubbing of dissimilar materials together—and from lightning, courtesy of B. Franklin. The glass separation between the tin foil stored the charge, much as condensers do today.

The early nineteenth century saw a veritable explosion of interest in the natural science of electricity. Franklin was only one of many inquisitive persons who experimented with electrical phenomena. While these men and women enjoyed their "shocking" experiences, scientists such as Michael Faraday were investigating what we know today as semiconductors. (Once again that remarkable experimenter Faraday must be credited with uncovering what today has become the basis of an entire industry.) In 1833 he observed that some materials became better or worse conductors of electricity at different temperatures. The notion that the conductivity of a wire could be varied and that some so-called conductors were better or worse than others gave birth to the semiconductor. Not long after, another experimenter explained just how and why the Leydon Jar worked, and how certain materials when joined together produced an electrical charge or current. This, the photovoltaic effect, was the beginning of our modern batteries.

Two additional discoveries, both made before 1875, laid the groundwork for today's semiconductor revolution. A gentleman named Smith found that when light was shone on a selenium chip, a voltage was created. This became known as the photo-optical effect. In 1874, Braun invented that wonderful tool used by the amateur radio buff—the cat's whisker, our first semiconductor device. The cat's whisker in the hands of the radio amateur looked amazingly like the early transistor—one or two wires connected to a base plate (see Figure 3-1). The challenge to the radio buff was to find the best spot on that base plate, the spot that would capture the weak radio signal.

Instead of amplifying the signal, as the transistor did, the cat's whisker rectified it. Rectification, or the ability to pass current in one direction only, is

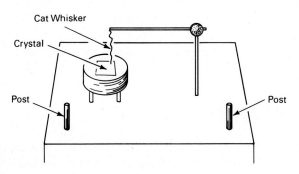

FIGURE 3-1
The Cat's Whisker: Radio's First Semiconductor. (From Jack Gould, *All About Radio and Television*, NY: Random House, 1953.)

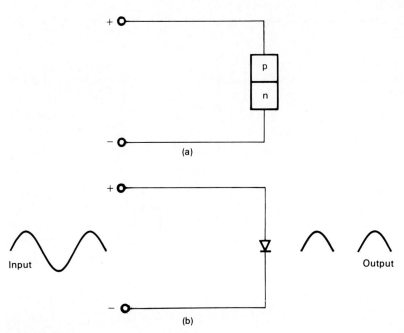

FIGURE 3-2
The Diode as a Semiconductor Rectifier.

essential to detecting signals. Turn for a moment to Figure 2-6. You will recall that when we selected a carrier for the long-distance transmission of signals we chose one that changed and moved in a regular manner so that we could predict just what it would do next. We selected the sinusoidal, a signal that alternates from a peak through a valley to another opposite peak—in short, an alternating signal. The information signal superimposed on the carrier in Figure 2-6 is irregular but does not flow in both directions. We want to separate or detect that information signal from the carrier. The cat's whisker does just that. A semiconductor rectifier known as a diode does that, too, as shown in Figure 3-2. Note that even though the current entering the diode moves in both directions, the current leaves in only one; it is *rectified*.

While physicists and chemists waxed enthusiastic about the wonders of the materials that were semiconducting, Bell Laboratories engineers recognized that the rectifier also could be used as a switch. If it passed current in only one direction, the switch would be closed when the current was in that direction and open when the current flowed in the opposite direction. The diode as a switch is illustrated in Figure 3-3.

MERVYN KELLEY'S WONDERFUL CHRISTMAS PRESENT

It is no surprise, then, that telephone company researchers were interested in this switch with no movable parts. As early as 1936 it had become quite evident

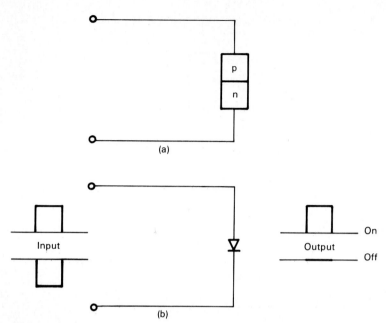

FIGURE 3-3
The Diode as a Switch.

to Mervyn Kelley, director of research at Bell Laboratories, that with the steady growth in telephone demand there soon would not be enough space to house the system's switches. Furthermore, these switches would require the power of a good-sized city to run. He wondered if it might be possible to reduce both the size and power requirements for telephone switching.

His fondest wishes were realized on the day before Christmas in 1947 when William Shockley, John Bardeen, and Walter Brattain connected two wires to a chip of germanium and invented the transistor. The era of the multitalented chip was born. Silicon has since replaced germanium but the principle of the transistor and of the integrated circuit it spawned has not changed over the years.

In its natural state, silicon does not readily conduct electricity. Each silicon atom has four electrons in its outermost shell. The atoms are arranged symmetrically in a solid crystal so that each atom is surrounded by a stable configuration of eight shared electrons, as shown in Figure 3-4(a). To turn silicon into a good semiconductor it must be "doped"—that is, another element must be introduced or diffused into the crystal structure to upset this stable configuration (see Figure 3-4(b)). Replace a few atoms of silicon with, say, atoms of phosphorus which are surrounded by five electrons and you create what is known as an n-type semiconductor. Replacing a silicon atom with a phosphorus atom introduces an "extra" electron, one which has no role to play in holding or

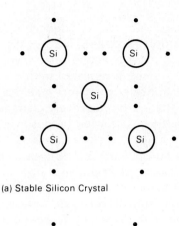

(a) Stable Silicon Crystal

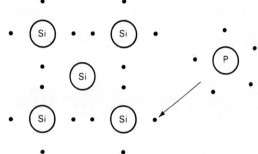

(b) "Doping" with Phosphorus

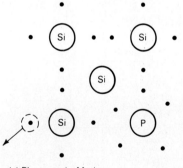

(c) Electrons in Motion

FIGURE 3-4
How to Create an "n-type" Semiconductor.

bonding the structure or lattice together and which can therefore be set into motion by applying a small voltage across the crystal, as illustrated in Figure 3-4(c). Electrons-in-motion are electrical current.

Alternatively, silicon can be doped with, say, boron to create a p-type semiconductor. Because a boron atom has only three electrons instead of the four in a silicon atom, the result of this substitution is an electron deficiency, a positively charged "hole" in the lattice. If you apply a voltage to the crystal, an electron from an adjacent atom in the lattice will move to fill the hole, leaving another hole behind to be filled in turn. The process is repeated so that, in effect, the hole is passed along from atom to atom. Electronics engineers say that current in an n-type semiconductor is carried by negatively charged electrons while current in a p-type semiconductor is carried by positively charged holes.

The principles of the transistor are illustrated in Figure 3-5. Suppose we make the voltage at the emitter zero and put a very small positive voltage on the base plate. Note that the path from the base to the emitter is our diode (see Figure 3-2), passing current in only one direction, from the base to the emitter, or a engineers say, holes flow from the positive base to the negative emitter. But with the presence of excess positive holes in the emitter, electrons will

FIGURE 3-5
How the Transistor Amplifies.

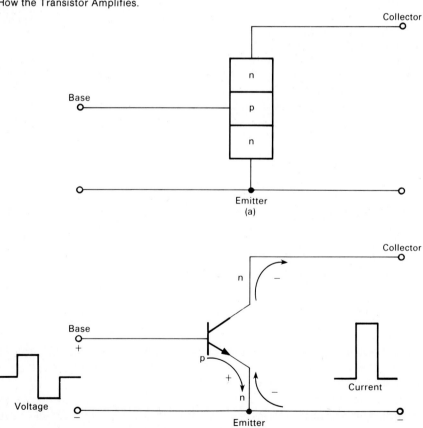

begin to flow into the emitter to counter the effect of the positive holes and from the emitter into the base. Where will these electrons go from here? They cannot flow back to the emitter so they must all flow into the collector. The result is a very large flow of current from the base into the collector.

We started with a very small voltage on the base and have ended up with a large current flow to the cathode; we have *amplified* the signal on the base. If we reduce the voltage on the base, no more holes will flow from base to emitter, no more electrons will flow from the emitter to the base, and no more electrons will flow into the collector. This is exactly what we want an amplifier to do—respond to small changes with large changes! Consequently, by changing the polarity on the base—that is, going from positive to negative—we can switch the current to the collector on or off. We have created another semiconductor switch and, what's more, a way of counting. Every time we raise the voltage on the base, we get a surge of current to the collector and if this current can be stored, we have closed the switch and counted a binary 1. When we reduce the current and turn the switch off we count a binary 0. It looks as if we have the beginnings of a computer.

CIRCUIT DESIGN WITH ONES AND ZEROS

In the middle of the nineteenth century the English philosopher George Boole wrote some papers on mathematical logic, a logic based on the notion that a statement is either true or false. Let us assign the number one to statements that are true and the number zero to those that are false. Going one step further, let's say that the transistor in Figure 3-5 delivers a one when it passes current and a zero when it does not. To put it another way, when the switch is closed it delivers information that means one or true; when the switch is open it delivers information that means zero or false.

In the previous chapter we saw that a digital signal is made up of a series of ones and zeros organized into a code that represents information. The telegraph produced these zeros and ones by sending a signal or not sending a signal. This was done by means of a relay that switched the current on or off.

Boole provided us with an algebraic means for representing this process of switching so that circuit designers can, 100 years later, determine how to create the chips that perform the intelligent functions in today's communications and computing systems.

It is one of the great virtues of digital technology that almost all of the functions required to send information can be performed by simply combining zeros and ones, by switching a current on or switching a current off. Today we use transistors as switches to translate Boole's true and false statements, switching current on or off, depending upon whether the statement is true (one) or false (zero). We use the transistors as gates to allow the flow or current for a one or to stop the flow of current for a zero.

The transistor described in Figure 3-5 is the fundamental unit of electronic digital circuitry. It is a gate, for, as we noted, it can either pass current or not pass current. Add another transistor and several passive components (resistors and capacitors) and we have one of the six basic building blocks of almost all digital functions. The one shown in Figure 3-6 is an *OR* gate. We show the gate in two ways, as a collection of components, (Figure 3-6(a)), and in the form of a switch (Figure 3-6(b)) which it really is. In the logic of philosopher Boole, this means that A plus B equals C or A *OR* B equals C. If switch A is closed, C equals A. If switches A and B are closed, C equals A and B. We are adding with either switches or transistor gates.

Five other gates are used to define logical functions which are transplanted into mathematical operations. One of these is an *AND* gate (shown in Figure 3-7). Again we provide it in its actual form as a collection of components (Figure 3-7(a)), or a circuit and as a set of switches (Figure 3-7(b)). In the logic of Boole, A times B equals C or A *AND* B equals C. If switch A is closed but switch B is open, C will equal zero. Similarly if switch B is open and A is open, C will equal zero. If both switches A and B are closed, then C equals AB. We are multiplying.

The five building blocks look remarkably similar as you can discover for yourself in the additional readings suggested at the end of this chapter. The wonder of it is that so few components so easily fabricated on tiny silicon chips can be combined in innumerable ways to create today's computers and telecommunication systems.

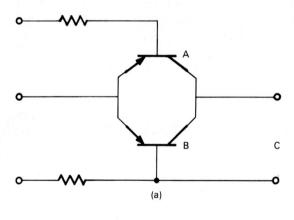

FIGURE 3-6
An *OR* Gate.

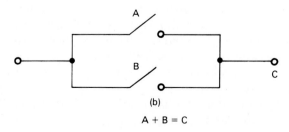

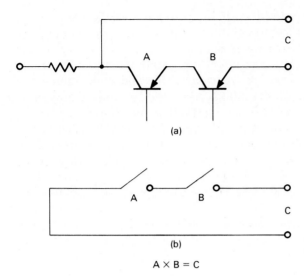

FIGURE 3-7
An *AND* Gate.

A × B = C

TAKING THE PEOPLE OUT OF THE PRODUCTION: THE INTEGRATED CIRCUIT

An electronic device is made up of active and passive circuit elements. Transistors, diodes, and the old-fashioned vacuum tube are active elements; resistors, capacitors, and inductors are passive elements.

It was once the universal practice to manufacture each component separately and then assemble the complete device by wiring these components together on a chassis or circuit board. Electronic functions have not changed; we still amplify, rectify, switch, convert, and perform all sorts of other actions necessary to communicate or compute, and the microelectronic devices are still made up of transistors, resistors, and capacitors. The difference is that all of these elements are fabricated on a single "plate," or substrate, and at the same time.

Depending on how you combine the n- and p-type regions—that is, as np, pn, npn, pnp—you can produce particular circuit functions, transistors that amplify, diodes that rectify or detect, and capacitors and resistors in various configurations. To complete the electrical circuit, you also produce the necessary connectors and insulation between functional devices.

The material upon which these devices are produced is usually a slice of silicon, about three or four inches in diameter and no more that 1/64 of an inch thick. The circuits are built up layer by layer selectively introducing the impurities needed to create the circuit functions desired. The secret of the low cost of silicon devices is in their manufacture, a process of mass production by chemical processing. Human intervention is minimized and thousands, if not millions, of these circuits can be made rapidly and cheaply. As in any other mass production, those circuits that do not meet specifications are simply rejected and the remaining material is used to start over again.

A somewhat more labor intensive, time consuming, and, consequently, more costly process involves laying out the circuit so that it is evident where and how the regions are placed on the base or substrate. Today, the pattern of each circuit layer is prepared by a computer programmed to lay out the circuit functions in the way the designer of the system wishes and, at the same time, conserve substrate material. The resulting artwork—the photolithographic mask—is usually about 250 times the size of the final microcircuit and is the precious output of long hours of design, the product of the information era that is highly valued and carefully protected by chip manufacturers. Each pattern layer is photographically reduced to the actual circuit (hardware) size and repeated over and over again to fill the entire area of the silicon slice. These become the photolithographic masks used to reproduce the patterns in "building up" the microcircuit.

These then are the manufacturing processes that are creating the information era and causing consternation throughout the world. These processes are doing away with the long benches of people who must insert components one-by-one and are thus replacing labor with packaged information.

Let us now put process and product together.

Once developed, the photolithographic mask is used to produce a silicon slice containing a series of identical microcircuits (the actual number depends on the functions the system is to perform). These functional modules, or building blocks, make up systems that perform signal processing; in other words, they are the input-output equipment or terminals for communicating that provide the intelligence needed in today's telecommunication systems.

RAM AND ROM—MEMORY FOR COMMUNICATIONS

We have talked so frequently about the convergence of computing with communicating it behooves us now to examine just what is meant. What precisely has computing brought to communicating that is so radically changing the nature of telecommunications? While we cannot point to a single factor and not be criticized for oversimplification, one important function the computer has brought to telecommunications stands out—the addition of memory.

Consider, for example, what the communication system must do in order to arrange an analog signal for time division multiplexing and to ensure that the receiver receives the message the sender has sent (see the section in Chapter 2 on pulse code modulation). A great deal of coordination is required so that the system remembers what it has been instructed to do. In subsequent chapters, especially in Chapter 7 on intelligent telephones, we shall see how important the ability to remember is to the transmission of information.

In the previous chapter we noted that there are some forms of communication in which the information is first recorded and stored for some time before it gets to the recipient. They include books, films, and paintings. Once stored in the memory—the pages, acetate strips, or canvas—of these communication forms, the information is not suitable for electronic transmission. But, if stored on magnetic tape or even in the grooves of vinyl records, the information can be recaptured for electronic transmission. Indeed, computers use several forms of bulk magnetic storage. The diskettes used by personal computers are magnetic storage devices, and, early in the history of computers, magnetic core memories were the primary forms of data storage.

When information is stored on a small magnet by means of a pulse (a binary zero, for example) impressed on the ferrite core, as the magnet was called, the core becomes magnetized in one direction. It retains that polarization until another pulse comes along. In other words, it remembers the information imparted to it by the first pulse.

When an electric charge was stored in the Leydon Jar, it, too "remembered" that charge until discharged into Franklin's guinea pigs or, in today's language of computers, until the stored information was read.

The microprocessors used in telecommunications devices, the stored program switches we shall encounter in Chapter 6 on the telephone, and the memories required to multiplex numerous digital messages along wires, cables, or over-the-air cannot often use these forms of "bulk" storage. They operate too slowly and use up too much space. Furthermore, they are expensive and one of the major objectives of using computer technology in telecommunications is to lower the cost of transmission. Consequently, we use semiconductors rather than magnetic memories for the high-speed performance required in our communications systems.

The transistor described in Figure 3-5 and the gates in Figures 3-6 and 3-7 are examples of such semiconductor memories. They remember the zero or one as long as we choose to let them remember—that is, until we write another zero or one in the memory or read the zero or one stored in it.

Semiconductor memory can be of several types. Read-only memory (ROM) cannot be altered or changed once the information is stored in the device. ROMs store information by arranging a pattern of connections between transistors on a chip, including a semiconductor storage capacitor, permanently. Programming a ROM requires changing the photolithographic mask used to fabricate the chip (see Figure 3-8).

Random-access memory (RAM) is used to hold data you might want to change after having read it. RAMs are volatile; if the power supply is cut off while using the chip, the information stored on it is lost (see Figure 3-9). Just as the ROM has a built-in semiconductor capacitor, so the RAM often has a built-in semiconductor magnet. RAMs are the memories that store data while you work—the information you may want to revise over and over again as you write with your personal computer or word processor. Because they are sort-term memories, they have often been called "scratch pad" memories. Just like your scratch pad, unless you transfer the information from the RAM to more permanent storage, (the magnetic disc in your computer or word processor or your notebook), this information will disappear when you turn off the computer or leave your desk.

In your travels through the computer-communications world you will come across PROM, a programmable-read-only memory in which the pattern of connections can be established after the circuit's manufacture. However, it is

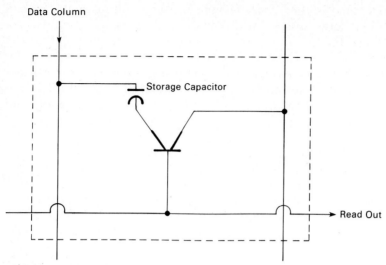

FIGURE 3-8
Read-Only Memory (the ROM).

still a rather permanent memory—once the pattern is established, it cannot be changed easily.

There are also erasable-programmable-read-only memories (EPROM) in which the stored program can be erased by exposing the chip to ultraviolet light, thanks to another wonderful attribute of semiconducting material discovered by nineteenth century-experimenters.

SUBSTITUTING SOFTWARE FOR HARDWARE

The mass production of chips is both a blessing and, sometimes, a bit of a curse. We have seen how the photochemical manufacture of chips can give us mass-produced circuits at very low cost. But this assumes that there is a great demand for the equipments and systems requiring these chips. Indeed, it is true that many household applications of chips, such as the microprocessors that control the cycles in washing machines and that store cooking instructions in microwave ovens, lead to the demand for large quantities of these chips. Consider, for example, the three-chip telephone shown in Figure 3-10. It is made up of a chip that connects the local telephone to the long-distance system, the line multiplexer or line mux, two identical filter chips that perform the frequency multiplexing function, and the codec, a device for converting the analog voice signal to a digital signal for transmission. Consider the number of telephones in the nation and you will see that this is both an excellent application for a mass produced chip and one that could be very low in cost.

Semiconductors for which there is great demand merit the high cost of facilities required for their production; the large number of chips produced

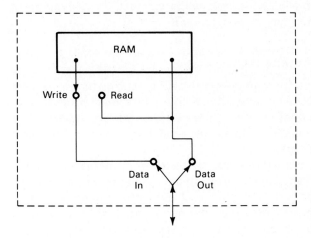

FIGURE 3-9
Random-Access Memory (the RAM).

quickly recoup this investment. But chips that are not much in demand, such as those for specialty circuits, will not benefit from the photolithographic chemical diffusion manufacturing process. The process does not change even if the number of chips produced is small and when limited quantities are produced, it can be costly. For this reason, designers seek to create even special functions from as many "standard" chips as possible.

One way to reduce the need for custom chips is to substitute programming for hardware. As noted in our discussion about "gates," a relatively small number of standard chips can be combined in many different ways to produce a much larger number of functions. This modular construction can be achieved either by "hardwiring" the design in the chip or by making the interconnection with software. By programming the instructions to the microprocessor that organize the activities of the chips, the same chip can be used over and over

FIGURE 3-10
A Three-Chip Telephone.

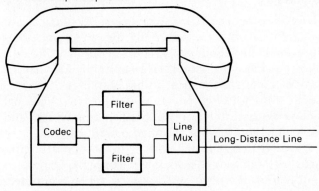

again. The cost of that chip will continue to fall as we make use of even more chips through clever instructions called *programs*.

A MOST FORTUITOUS MARRIAGE

The convergence of computing and communicating about which you have heard so much would not have been possible without the semiconductor chip, a most appropriate technology for computer-communications systems. It is not often that such a fortunate marriage takes place but these are the events that make innovations possible.

Semiconductor technologies are used to perform an increasingly wide variety of functions in the transmission of information. They make possible the multiplexing techniques discussed in Chapter 2 and without them we would not be able to sample information rapidly enough to enable us to send multiple messages over the same transmission space. Semiconductors are the switches of today's electronic exchanges in the central offices of telephone companies or in modern office switchboards, the intelligent Private Automatic Branch Exchanges (PABX) discussed in Chapter 6. These systems make possible the many new telephone and data services, including electronic mail, teleconferencing, and others to be examined in Chapter 7 on intelligent telephones. Clearly, semiconductors are not only multitalented, they can be quite intelligent. They provide the memory, control, and coordination needed to manage and manipulate the assemblage of equipment and software. They make up the microprocessors of computers and telecommunication systems. Semiconducting devices link the users to the telecommunications systems and computers; they perform the analog-to-digital and digital-to-analog conversions that enable people to communicate in bits; they are the workings of the input-output devices or terminals we discuss in Chapter 5. Had the chip been available when Bell was experimenting with the harmonic telegraph and the telephone, we might have gone "digital" a hundred years ago. Remember, the telegraph is a digital system.

THE MANY CHIPS IN OUR LIVES

In order to amortize high development costs, manufacturers seek applications which demand millions of identical units. Consequently, the semiconductor is rapidly becoming omnipresent in just about every aspect of our lives. There is hardly a consumer product that does not contain at least one chip. Calculators, video games, and home appliances are the most visible examples of devices made possible by the semiconductor chip. Not so visible are the chips that control heating and cooling systems, telephones, electric typewriters, and personal computers. Even television sets for those who want the luxury of remote tuning by the light touch of a button in sleepy hands incorporate chips.

Automobile dashboards are beginning to resemble the cockpits of modern jets with engine efficiency, touring speed settings, and other computer-semiconductor displays of controlled measures designed to reduce fuel consumption. Semiconductors have found their way into the direct operation of the automobile engine through electronic or computer-controlled fuel injection.

Compared to traditional telecommunications equipment technologies, semiconductors are more reliable and use less power and less space, just as Mervyn Kelley desired. They are also less expensive, a factor for which Mr. Kelley would not have dared to hope.

We can expect that semiconductors will be used increasingly to provide improved links between man and machine; it will be semiconducting devices that make the talking typewriter a reality. It will be semiconducting devices that make ever more efficient the use of valuable communications bandwidth and thereby reduce the cost of communications transmission. And it will be semiconductors that remember all of the instructions we give our telephone to wake us up in the morning, and to remember the last number dialed and the numbers must frequently dialed.

The minimal limits of the semiconductor chip's size and cost have not been reached. Nor have all of its applications been identified. The number of components on a chip continues to double every year; more and more functions are created on chips that are growing smaller. By 1990, the price per operation or per instruction will decline by at least two orders of magnitude—that is, the cost to tell a switch to turn on will be about 1/1000 of a penny. The more chips manufactured, the better and cheaper they become (see Figure 3-11). The same is true of memories; we learned how to make memories with

FIGURE 3-11
Experience Drives the Price of Semiconductors Down.

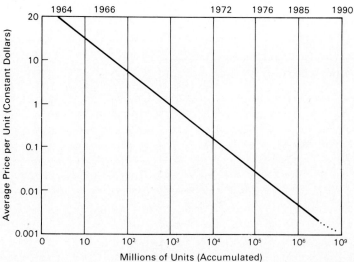

16,000 bits, and 64,000 bits; now the 256,000 bit barrier has been broken and we are moving toward the 1,000,000 bit chip. The price of the 1M bit memory chip in 1990 is not known but one thing is certain, it will be low. It is clear, too, that the cost per bit of computer memory—the cost of storing a yes or a no—is rapidly falling to about the same as the cost for switch instruction (see Figure 3-12).

Think of the chip's functions as actions a human operator might have to perform. For example, consider the clerk at the checkout counter who must remember or look up prices of, say, 100 grocery items. If each item requires three digits, the clerk must remember or look into 300 yes or no instructions—600 bits—every time an item is purchased.

The cost of the semiconductor memory required to store this information could be less that 6/10 of a penny. Compare this to the cost of printing daily price lists, the pricing errors that would most certainly occur, and the unhappy customers who must wait for corrections or who find they have been overcharged.

The number of functions on chips and the cost of semiconductor devices changed, not by fractions, but by orders of magnitude over the years. In this regard, we talk about the scale of integration. During the past ten years we have gone from packaging about 75,000 components on a chip—a process called large scale integration (LSI)—to well over 200,000 components, which is called very large scale integration (VLSI). What do you suppose the next scale of integration will be called?

IS THERE A POSTCHIP SOCIETY?

The information society, or the information era, is seen increasingly as the result of the computer and, even more specifically, of the semiconductor chip.

FIGURE 3-12
The Rapidly Falling Cost of Memory.

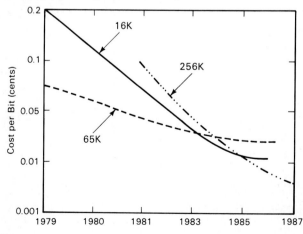

Economists, politicians, sociologists, engineers, and philosophers point to this remarkable device as the primary actor on the world's industry stage today, and they are seeking explanations and guidance on how to cope with a postchip society. Indeed, it is difficult to understand the ramifications of this remarkable little device with its power to act intelligently, a characteristic once thought to be safely in human hands.

It appears that remembering the information needed to respond to more than 100 questions can be accomplished by a chip and some associated hardware that together might well cost no more than $25. Not too long ago a line of fifty assemblers each inserting, say, ten components such as vacuum tubes or transistors, resistors, capacitors, and inductance coils into a chassis would have been required to assemble a television set. Today there may be less than half that number handling a much smaller number of chips containing the equivalent tubes, transistors, coils, resistors, capacitors, and wiring that make up the television set. Each chip replaces not only the electrical components but the workers required to solder them into place.

Realistically, various factors will limit the capability of the industry to continue increasing the number of functions on a chip. As the number of functions per chip increases there will be concern over the quality of the silicon or other material being used. As pointed out earlier in this chapter, circuit functions are created by careful "doping" of the base material; if the chip does not contain pure material it is impossible to control the creation of circuit functions.

In addition, as functions are packed more densely, the wires that interconnect these functions become smaller and smaller. If these interconnections become as small as the wavelength of light it becomes difficult, if not entirely impossible, for the camera to pick out these very tiny wires.

There are also limitations on the speed at which electrons can move between functions. The resistance in the interconnections themselves limits this speed, hence the development of the Josephson junction in which circuits are operated at temperatures so low the resistance of wires almost disappears.

Along with these physical and chemical limitations there are improved manufacturing equipment and techniques required in order to stay ahead of the competition. Firms in almost every developed and developing nation see the chip as their entree into the information society.

ADDITIONAL READINGS

The semiconductor can be seen as an extremely useful design tool for engineers or as a remarkable instrument of social change. It is both. We suggest readings that will enable you to explore both views as well as the many in between.

Braun, Ernest, and Stuart Macdonald, *Revolution in Miniature* 3rd ed. (New York: Cambridge University Press, 1982). An historical view of the microelectronics revolution with insights into its consequences.

Evans, Christopher, *The Micro Millenium,* (New York: The Viking Press 1979). A popularly and very optimistic written account of the wonders of the semiconductor.

Forestor, Tom, ed., *The Microelectronic Revolution* (Cambridge, MA: MIT Press, 1981). A most comprehensive selection of excellent papers dealing with just about every aspect of microelectronics.

Pierce, John R., *Signals: The Telephone and Beyond* (San Francisco: W.H. Freeman, 1981). This small book provides an excellent introduction to how transistors work in switching, as well as other aspects of information and its transmission by telephone.

Scientific American, *Microelectronics* (San Francisco: W. H. Freeman, 1977). A collection of technology-oriented papers on microelectronics that takes into consideration the circuits and hardware.

CHAPTER 4

THE COMMUNICATIONS SATELLITE: NEWTON AND CLARKE COOPERATE

Two technological innovations, the semiconductor chip and the communications satellite, are responsible for the emerging information age. The previous chapter discussed the multitalented chip; now we turn our attention to the satellite.

The satellite has made communications possible where previously it was either impossible or very difficult. The satellite knows no national boundaries; it deals with the nations of the world as if they were one. The communications satellite has captured the imagination of all nations, advanced and underdeveloped; it has expanded visions as no religion has in modern times and it has encouraged hopes for human betterment in a world where hope has often been absent. Some say that the satellite will democratize the world and bring all peoples together in a world "society." In this "one world," information and its communication will be a prerequisite for survival. The satellite will make this information available to everyone immediately and simultaneously, thus it is viewed as presenting immense opportunities as well as threats that are yet unknown.

SATELLITES AS EXTRATERRESTRIAL RELAYS

In his famous "Wireless World" paper of October 1945, Arthur C. Clarke of *2001: A Space Odyssey* fame suggested that "a true broadcast service giving constant field strength at all times over the whole of the globe would be invaluable, not to say, indispensable, in a world society." He proposed that "rockets" in suitable extraterrestrial positions (as shown in Figure 4-1) could overcome the difficulties with high-frequency radio transmissions and the "peculiarities of the ionosphere," and make feasible not only worldwide radio and telephone but also television. Thus was born the notion of satellite communications.

Today much if not all of this nation's television is delivered to network broadcast stations by satellite for retransmission, either over-the-air or via cable, to home receivers. Without satellite transmission, the current boom in pay-TV and the resulting rapid expansion of cable television itself would have been impossible. It was the use of the satellite by Home Box Office to deliver its programming to client cable systems in about 1977 that precipitated the explosion in cable television. Without the satellite, such programming would have required the delivery of cassettes to cable systems by mail or even messenger (a process broadcasters call "bicycling,") with all the attendant uncertainties of cassette loss, delay, or even theft. Another alternative might have been the purchase of very expensive terrestrial microwave carriers from Ma Bell. (We will discuss these line-haul services in Chapter 6.)

Now, a new class of satellites is on the launch pads, greater in power and with the ability to "focus" a signal. These satellites will be able to provide television broadcasting directly to antennae mounted on rooftops. This emerging video option promises to deliver multiple channels of television, pay-TV or broadcast TV, with and without commercials, directly to the home. Small, low-cost (perhaps in the neighborhood of $300 or less) antennae are also in the works. What this will do to our broadcast systems is a hot topic of debate which we shall examine in our chapter on broadcast and cable systems (Chapter 9).

Satellites are not solely an American phenomenon. The Europeans also have been planning for a new era in telecommunications and, of course, broadcasting. Now that the initial tests of the Orbital Test Satellite launched by the European Space Agency have been completed, plans are under way for covering much of the continent, from the frontiers of Ireland to the Polish-Russian

FIGURE 4-1
Arthur C. Clarke's "Rockets" in Suitable Extraterrestrial Positions. (From "Extraterrestrial Relays; Can Rocket Stations Give Worldwide Radio Coverage?" *Wireless World*, Vol. LI, January–December, 1945 (with permission).)

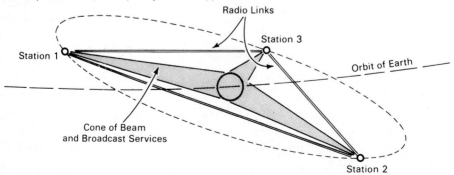

border, with signals "focused" for pickup by small antennae, perhaps in the neighborhood of six to nine feet in diameter. Needless to say, there is a great deal of political concern centered on this potential. The system's saving grace is that most households will not be able to afford necessary "dishes" yet, and broadcast stations can control retransmission. One of the reasons for the rapid growth in cable television in some European nations such as Belgium, the Netherlands, and the small principalities of Lichtenstein and Luxembourg is certainly the expectation of more satellite television from other nations.

The manner in which satellite technology is developing seems to ensure that the receiving antennae will get smaller and cheaper, more in line with what communities and even households can afford. This is all to the good for satellites offer special benefits for small countries. The Austrian National Broadcasting Organization has argued that by the end of the 1980s between ten and eighteen satellite television programs will be available to its viewers, a great deal more than they could afford otherwise. While every national television network will lose viewers to foreign stations, this loss will surely be compensated by their own gain in viewers from outside nations. Indeed, each nation is likely to be able to reach far more viewers than are available within its own boundaries. What a boon for advertisers!

Take a look now at some numbers and Figure 4-2. If Luxembourg were to launch its own satellite (as it has threatened to and could since it has long operated a commercial radio station within the nationalized broadcast borders of France and Germany), its signal could reach more than 200 million people in East and West Germany, France, Italy, Austria, Yugoslavia, Czechoslovakia, Hungary, and Poland with parts of Great Britain and Spain thrown in for good measure. Consider the problems faced by politicians who want to protect their national cultures against those who want to export theirs.

Indeed, broadcast satellites do not belong only to advanced nations. Developing nations are finding that the satellite enables them to leap directly into the information age with all its consequences, good and bad. India launched a satellite program not only to allow its industries and businesses to communicate with the world, but also as a way of making its diverse peoples aware of nationhood and to motivate and educate them for participation in national development.

Indonesia, with its more than 155 million people scattered over some 4000 islands stretching well beyond 2400 square miles, had the same idea when it launched the Palapa Satellite in 1977. What better way to communicate when roads can only carry you across a single island?

Reflect for a moment on the effect the broadcast satellite is having on the world. For years Europe dreamed of and struggled over the idea of a United States of Europe. Certainly the pressures of competition from the United States and Japan can be given much of the credit for drawing the European nations together. However, we cannot overlook the effect of the satellite with its ability to communicate, entertain, and advertise across national boundaries and cultures. Whether television homogenizes cultures or allows them to

FIGURE 4-2
Luxembourg Launches a Satellite.

expand and flourish might well be determined over the next several decades in Europe and in Southeast Asia where Palapa has become the regional voice of ASEAN, of Thailand, Indonesia, the Philippines, Singapore, Malaysia, and Brunei.

Satellites are not devices for television transmission only. Indeed, INTELSAT, the worldwide satellite carrier, is used primarily for the delivery of telephone voice service and for data transmission. As interest in digital transmission grows so grows the use of satellites for sending information in the form of data. This is because the satellite offers that wonderful combination of wide bandwidth and costs that are insensitive to distance. Let's discuss why.

THE IDEA OF A SATELLITE

Sir Isaac Newton hypothesized that if he could climb a high enough mountain and shoot a cannonball into the air, the ball would fall toward the earth at the same rate as the earth curves away from him (see Figure 4-3). The cannonball would thus continue to circle the earth without ever falling to earth; it would be a manmade moon around the earth or, what we today call, a satellite.

Newton never found his mountain and even if he had, the earth's atmosphere would have slowed the cannonball so much that it would have soon crashed to

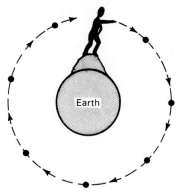

FIGURE 4-3
Isaac Newton and His Cannonball Create an Artificial Moon.

the earth. But if Newton could have shot the cannonball out of the earth's atmosphere at the speed of eight kilometers or five miles per second, it would never have returned and it would circle the earth indefinitely with no expenditure of power. It would become, just as Newton predicted, a second moon.

When a satellite circles the earth in this way it is in a *stable orbit*. An infinite number of possible stable orbits exists. An escape speed or orbital velocity of five miles per second can lift a satellite to a stable orbit closest to the earth, just outside the atmosphere, where it will circle the globe once every ninety minutes. The satellite will be in an elliptical orbit at an apogee (the highest point) of 300 miles and a perigee (the lowest point) of 100 miles as shown in Figure 4-4(a). To use this satellite for communications you would have to be able to see it, for, as noted in Chapter 2, radio waves, especially the high frequencies used in communicating with satellites, travel in straight lines. But as the earth moves, you would be able to see this satellite for only about a quarter of an hour, not even enough time for a single episode of M*A*S*H.

Suppose we increase the velocity at which the satellite leaves the earth (by giving it another boost once it is in its first stable orbit) and thus increase the radius of the orbit. The velocity at which the satellite circles the earth now decreases since gravity is less and the vehicle needs less centrifugal force to balance the lesser pull of the earth. Let's put our satellite into an orbit with radii of, say 6000 and 12,000 miles from the earth as shown in Figure 4-4(b). The satellite circles the earth every five to twelve hours and we can see it from anyplace on earth for approximately two hours at a time—just about right for a two-hour TV special. However, if we were to send the satellite out to 22,300 miles, it would be seen all the time and although it would appear to be stationary, it would really be rotating at the same speed as the earth (see Figure 4-4(c)). This is know as *geosynchronous orbit* and is a feature of today's communications satellites.

Since there are an infinite number of stable orbits for satellites, not every satellite must roam around the earth in circular orbit. There is no reason why a satellite can't have an elliptical orbit as in Figure 4-4(b). After all, the planets in our solar system move in elliptical orbits and do quite well insofar as they do

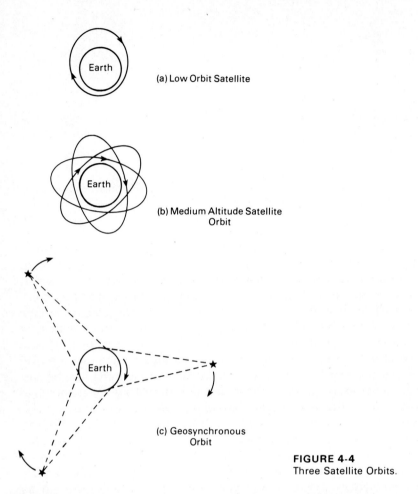

FIGURE 4-4
Three Satellite Orbits.

not fall into the sun. Indeed, most of the satellites in space today are in elliptical orbits, especially those satellites used by the military for surveillance and specialized communications. The satellites used for mapping, for surveying earth resources, and for identifying weather conditions perform better in elliptical orbits because they need to be closer to the earth at certain times. It makes sense for the Soviet Molniya communications satellite to be in an elliptical orbit in order to cover the ellipse-like shape of that vast country.

THE TECHNOLOGY OF SATELLITES

Now that we know Isaac Newton was right, let's see how Arthur Clarke would communicate with satellites.

A satellite is quite simple in function but extremely complex in performance. Functionally, it is the textbook parallel of the Shannon-Weaver linear communication model (shown in Figure 4-5) that today's communications textbooks

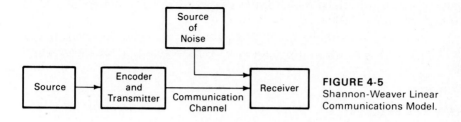

FIGURE 4-5
Shannon-Weaver Linear Communications Model.

love to display. The model contains a transmitter, receiver, distribution channel, and noise in the system. What makes this model so useful is that satellite businesses, services, and engineering can be conveniently divided up in the same manner. In Figure 4-6, we can see there is the space segment, or the satellite itself; the earth segment, our distribution channel where we fit the transmitting and receiving functions; and the noise in the channel, the various electromagnetic and planetary interferences to the signals traversing the 44,000 and more miles from transmitting to receiving locations by way of the satellite.

The complexity in satellite communication systems is created by engineering rather than theory. It is an expensive task to launch the vehicle into the desired orbit, although the space shuttle is offering bargain rates. And it requires complex engineering to keep the vehicle in its proper position in space for capturing the sun's rays for power, while maintaining precise focus on the ground stations in order to allow the weak signals carrying the information this system is designed to transmit to be captured by the receiving antennae. It is also a difficult engineering task to select and maintain frequencies in order to eliminate interference from the crowded electromagnetic traffic on the ground. There is the added concern that all of this carefully designed and adjusted

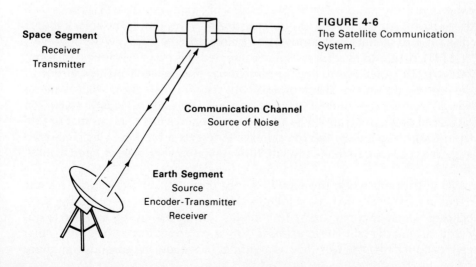

FIGURE 4-6
The Satellite Communication System.

equipment must withstand the enormous shock of being launched into space at speeds of up to 25,000 miles per hour.

We shall examine satellite technology not as engineers, however, but as intelligent users of its communications and as participants in the information age.

THE SPACE SEGMENT

The space segment of the satellite system is the spacecraft or vehicle itself. The spacecraft contains five major subsystems: 1) its power source; 2) its positioning and orientation devices; 3) communications for controlling the vehicle in orbit—what the engineers call the telemetry systems; 4) antennae for receiving and transmitting signals; and 5) the all-important transponders for processing and delivering the signals between earth stations.

The Power Source

Satellites are powered by solar cells—devices which have the ability to convert the sun's rays into electricity. These solar cells are devices similar to those discussed in Chapter 3; they are usually made of doped silicon crystals. To get the most power out of the solar cells it is necessary to expose a great deal of area to the sun. Consequently, the cells are arranged in panels on the outer surface of the vehicle, a configuration which essentially makes up the outer layer of the satellite structure. The solar cells are quite small, often no more than two centimeters square, and a great many of them are needed to generate a reasonable amount of power. They are not terribly efficient; indeed solar cells lose more than 60 percent of the sun's energy in the process of converting that energy into power to operate the satellite's communicating functions.

The more power the cells can generate, the more power the satellite can deliver to the ground stations 22,300 miles away, and the smaller and less expensive these stations can be. Since there are many more earth stations than there are orbiting vehicles, low cost is extremely desirable. The solar cells for the Early Bird satellite launched in 1965 delivered a mere forty watts of power; WESTAR in 1976 used solar cells that increased this power six times, to about 240 watts. The satellites circling the earth today that deliver a healthy portion of the world's international telephone conversations use about 1000 watts of power; some experimental satellites even go as high as 12,000 watts. The increased power is achieved by all sorts of ingenious mechanisms that unfold large wings or panels when the satellite enters its orbit. The wings, covered with solar cells, often make the satellites appear as very large winged insects.

What happens when the satellite is not directly facing the sun? This can occur at certain periods in a satellite's orbit around the earth or during solar eclipses which occur on forty-four nights during the spring and forty-four nights during the fall. These eclipses do not last long; but without sun, solar cells cannot deliver power. To compensate for the loss of power during these blackout periods, several nickel-cadmium storage batteries are on board. The lives of these batteries are very important to the life of the satellite; new developments in batteries are constantly increasing their longevity and, hence, the longevity of the satellite.

The Spacecraft Positioning and Orientation System

Arthur Clarke was not only a good and practical mathematician, he had imagination. Although well aware of Isaac Newton's infinite number of possible satellite orbits, he recognized that the worldwide coverage his communications satellite could deliver would best be achieved if the artificial "moon" circled the earth above the earth's equator. Consequently, communications satellites are positioned on a very narrow band that circles the earth 22,300 miles above the equator.

The early satellites had limited power and needed very large antennae—that is, devices to pick up signals from the ground and transmit signals to the ground (see Chapter 2). They also needed quite a bit of space between them in order to avoid any electromagnetic interference between the signals they were sending and receiving. They were spaced five degrees apart (or roughly 2700 miles from each other). This did not leave much electromagnetic interference-free room for many satellites; at best, only seventy-two satellites could share this very valuable real estate. In the early stages of satellite communications, many people did not believe there would be a great need for satellites for long-distance transmission because there were many transoceanic cables in place and several more planned. It wasn't long, however, before nations realized the social potential of the satellite and their dreams began in earnest. Almost every nation—large, small, developing, and developed—wanted its own space in the geosynchronous orbit, and the demand for space soon outgrew the availability of orbital slots.

It became apparent, too, that not every location on this orbit was equally valuable. What is the commercial value of providing communications to the wide and empty spaces of the Pacific Ocean with its small and underpopulated islands or to the yet-to-be developed nations of Africa where the need for sophisticated communications was only just beginning to be recognized? Keep in mind that delivering communications by satellite is a most expensive undertaking whether it is paid for by governments or private investors.

Fortunately for all, technology came to the rescue again. Improved electronics, better power sources, and more experience in launching the vehicles resulted in a reduction of the spacing required—first to four degrees, then to three degrees and now as close as two degrees (about 1100 miles apart). The distance necessary between satellites depends on their operating frequencies, the bandwidth of the transmitting stations, and on the satellites used—in short, on the electromagnetic distance rather than the number of miles. We shall examine the uses and transmitting bandwidths in our discussion of the satellite system's ground segment.

The narrow belt circling the earth at the equator is becoming very crowded, as you can see in Figures 4-7(a) and (b), and is even more crowded on that part of the belt serving North America. (See Figure 4-8.)

Once in place in that relatively small piece of real estate, the orbiting vehicle must maintain its position, not only with respect to neighboring vehicles (not because they will bump into each other but because of electromagnetic interference from their antennae), but in relation to the sun (the satellite's solar panels must face the sun and its antennae focus on the desired target—footprint—area on the ground).

The positioning of the satellite in space is controlled by a "despun stabilizing mechanism" and devices that emit short jet streams of gas to correct any shifting or rolling the vehicle might do while in orbit. The stabilizing mechanism is a mechanical one, requiring little power once the spinning has begun—after all, there is no gravity or air resistance at 22,300 miles in space. The satellites, minus the antennae, are spun around on their axes in one direction while their antennae rotate in the opposite direction at the same rate. On some "birds," internal momentum wheels are used to maintain stability. Once again Newton comes to the rescue with his Third Law—action equals reaction—which showed the engineers how to stabilize the vehicle on the equatorial orbit.

The gas jets do, however, require some power in the form of the gas that is stored in the vehicle when it is launched. If the satellite drifts around too much due to the gravity effects of nearby satellites, the gas will be used up. This, then, is the second of the two factors that limit the life of the satellite, the first being the life of the nickel-cadmium solar batteries discussed previously.

The Control and Operational Telemetry

These controlling actions may be initiated automatically in the satellite itself, using the sensing and control systems on board the vehicle. The vehicle is also under constant surveillance by operators and controlling systems on the ground. They monitor sensors that locate the satellite's position with respect to the sun, the one truly stable body in our solar system. These sensors radio an indication of the vehicle's relative position to control stations which respond by sending back correcting signals. The signals turn on the gas jets and, if necessary, impart some spin to the spinning mechanisms we described previously. Engineers call this monitoring of signals that inform them about the health of remotely operating electric power stations, monitor robots on the factory floor, or patients in a hospital intensive care facility *telemetering*.

Accurately positioning satellites in space is critical to their ability to deliver information to the desired location. Increasingly, it is necessary to focus the signals more and more narrowly. Unlike the wide angles of satellite coverage shown in Figure 4-2, we now want to deliver signals to a specific city or even a specific neighborhood using "spot" beams. This requires that the satellite keep its position steady to as little variation as plus or minus one degree.

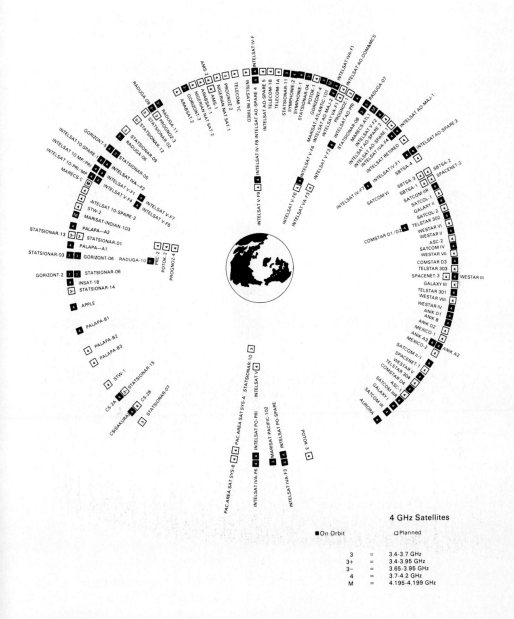

FIGURE 4-7(a)
Synchronous Satellite Equatorial Parking Orbit (C-Band).

Prepared for Satellite Communications by Walter L. Morgan, Communications Center of Clarksburg, (301) 428-9000, November, 1983 (with permission).

FIGURE 4-7(b)
Synchronous Satellite Equatorial Parking Orbit (K-Band).

11 & 12 GHz Satellites
FSS

■ On Orbit 11/1/83   □ Planned
                      ● BSS (Planned)

11	=	10.95-11.2, 11.45-11.7 GHz
12	=	11.7-12.2 GHz
12+	=	11.7-12.5 GHz
13–	=	12.2-12.7 GHz
13	=	13.4-13.7 GHz
14 only	=	uplink at 14 GHz downlink at another frequency

Prepared for Satellite Communications by Walter L. Morgan Communications Center of Clarksburg, (301) 428-9000, November, 1983 (with permission).

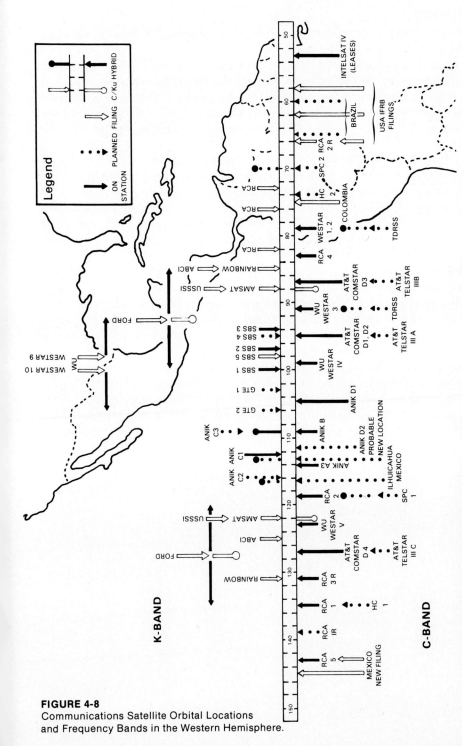

FIGURE 4-8
Communications Satellite Orbital Locations and Frequency Bands in the Western Hemisphere.

The Antennae

The devices that perform the remarkable job of focusing beams of all sizes to a precise location are the antennae.

In Chapter 2 we described how electromagnetic waves are transmitted or broadcast through the atmosphere. In that brief description we discussed the wires through which current passes and thereby generates the electromagnetic field we call the radio wave. There are also wires that receive these waves. They are antennae for transmission and reception of broadcast signals in their simplest forms.

It is possible to "shape" the electromagnetic wave transmitted in patterns or figures that are best suited to the desired area to be covered by the broadcast signal. Designing antennae that can accommodate the variety of services and area coverages satellite customers want is a highly specialized art, in many ways similar to the design of optical mirrors and reflectors. Indeed, the language of antenna design is sprinkled with such terms as aperture, shaping of beams, reflector, and polarization.

Antennae are designed to deliver signals with sufficient power to be detected in areas where there is usually a great deal of electromagnetic interference, such as in space where there may be neighboring satellites transmitting signals that will cause interference and where there is the galactic noise we described in Chapter 2. How well the antenna can be designed to accomplish all of this really determines what the satellite can deliver.

Arthur Clarke's dream of worldwide radio coverage required three antennae to deliver signals that would each cover a third of the earth as shown in Figure 4-9. These are the *footprints* of INTELSAT, the international satellite carrier that serves the world. These footprints deliver global beams by means of satellites over the Atlantic, Pacific, and Indian oceans (Figure 4-10). As we might expect

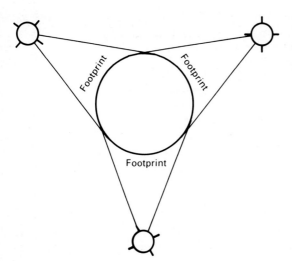

FIGURE 4-9
The Clarke Orbits and Their Footprints.

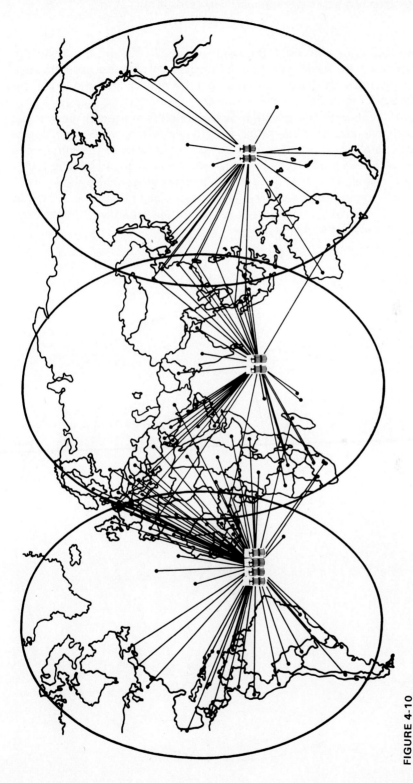

FIGURE 4-10
The Intelsat Global System.

Source: CONSAT, Communications Satellite Corporation Magazine, Nov. 14, 1984, (with permission).

from our experience with light, the larger the area covered, the lower the intensity or power delivered to the area. Consequently, it is necessary to have very large antennae on the ground to detect the signals; and, the larger the ground systems, the more expensive they are.

INTELSAT deals with the problem by providing footprints or beams that cover less than a third of the earth, say only the eastern or western portion of the global beam as shown in Figure 4-11. As the intensity of the signal—the power delivered—is greater, the antennae can be smaller and cheaper.

But that's not all. INTELSAT (and other satellites as well) also deliver spot beams which cover comparatively small portions of the earth as shown in Figure 4-12. The power delivered is greater and the antenna system on the ground can be smaller and cheaper.

FIGURE 4-11
Intelsat Hemisphere Beam.

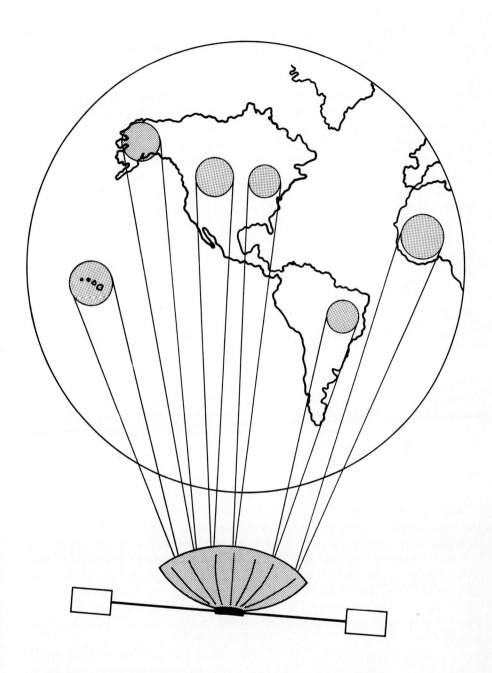

FIGURE 4-12
Spot Beams: How NASA's ATS-6 Pinpoints the Earth.

Clearly, there are tradeoffs designers can make between the size and power in the vehicle and the area to be covered on the ground. But there are other important parameters that can be traded off against one another; indeed, we shall encounter design tradeoffs throughout all of telecommunications and they are best demonstrated in the business of satellite systems design.

A single satellite can deliver multiple beams, including several spot beams, to a target area. No wonder these vehicles are getting larger and heavier and require increasingly greater power. This is why the concept of a space station or platform with many antennae in one location, on one satellite is becoming more attractive. In addition, with fewer good locations available on the equatorial orbit, it is advantageous to put as much transmitting capability as possible at a single location. That is why the space shuttle with its ability to carry very large and heavy "payloads" into a relatively low earth orbit may soon become the preferred way of launching satellites. Once in that "parking" orbit, the communications equipment can be lifted to a permanent position in a geosynchronous equatorial slot with less thrust.

The Transponders

Transporting information across continents, communicating between places where heretofore it was difficult, if not impossible, is what the satellite was designed to do. All that we have discussed so far is getting the communications hardware into the position where it can do the job of communicating, of transporting information from one part of the world to another cheaper and better than can be done with cable or other ground-based telecommunications systems. In the satellite, these all-important communication tasks are performed by transponders, or repeaters.

When all is finally said and done, the satellite is a very sophisticated repeater—that is, a system for receiving information from one place and sending it to another without in any way changing the information. It simply repeats the information by transmitting the same signal it has received to another location. Upcoming chapters on the telephone and on broadcasting emphasize the importance of this repeating function in all telecommunications systems. We have been performing this function for years—but not in space. And we don't use the same equipment in space as we would use on the ground. After all, as long as you are going to use multimillions of dollars to get these repeaters into equatorial orbit 22,300 miles in space, you want them to do the repeating job very efficiently; you want to get the most "bang for your buck."

The transponder is where all the information "action" takes place in the spacecraft. The mechanism is a highly complex electronic system operating at very high frequencies, in the gigahertz range—1000 million cycles per second (GHz)—and at high power. It uses the traveling wave tube or TWT. In this device the signal to be amplified travels down the tube in a helical pattern while a beam of electrons is fired from behind the small signal, bunching it and causing another signal to begin down the tube. The two signals—the original one and the new bunched signal—join together causing more bunching and more joining so that the original signal is greatly amplified. The longer the tube, the more the signal is amplified and the higher the power. Small wonder these TWTs need to be water cooled even in space.

The transponder receives signals from the transmitting earth station—signals that are very weak after their 22,300-mile trip into space—amplifies these signals, and retransmits them 22,300 miles to receivers on earth. The receiving and transmitting frequencies must be different, otherwise they would interfere with one another. The International Telecommunications Union (ITU), the international body that sets standards for communications, assigns frequencies for these links; for the *uplink,* ITU has assigned the frequency range of 5.95 to 6.45GHz and for the *downlink,* 3.7 to 4.2GHz. These are the frequencies you frequently hear referred to as the 4/6GHz or C-band.

Because space is an electromagnetically noisy place, cities are becoming increasingly jammed with all sorts of radio signals, and we want to cram as many vehicles as we can into that valuable but limited equatorial orbit, satellites are designed to operate at different frequencies so they will not interfere with one another. New satellites are designed to operate at 12/14GHz in the K-band (uplink/downlink). Plans are under way to use even higher frequencies, 20/30GHz and 50/60GHz, even though there would be degradation of the signal due to rain droplets in the air and more power would be required to make up for the losses.

Transponders can be viewed as "electronic real estate." Indeed, the recently launched Galaxy family of satellites is often referred to as "condominium" satellites. A satellite user will purchase or "rent" a transponder and put its bandwidth to the most cost-effective use as channels or circuits for voice or two-way video. INTELSAT deals in half circuits which are just what they seem—one-half of a channel or circuit. To a great extent this terminology is an accounting artifact of use to INTELSAT which must account to its many shareholders as to who has used what, for how long, and at what cost, in order to determine the distribution of profits.

Most transponders have a bandwidth of 36 megahertz (MHz) or some multiple thereof, such as 72MHz, although there are some variations in the satellites now being designed. Transponder bandwidth depends on the amount of traffic the buyer or renter wants to transmit. If, for example, the earth station is used as a common carrier—that is, available to and used by a great many firms—there will be large amounts of traffic; voice, data, or video and larger bandwidth transponders will be required. However, this depends a great deal on the nature of the earth-based portion of the satellite system, as we shall see in the next section of this chapter.

For purposes of discussion here, let us use the 36MHz transponder. While there are usually twelve of these transponders on a vehicle, because they are reused by means of *polarization,* we usually account for twenty-four of them. Simply adding up the number of transponders and multiplying that number by bandwidth tells us that a satellite has a bandwidth of about 800 MHz. That is not the whole story, however. How we use this bandwidth—what engineers call the effective usable bandwidth—determines its true value. The use of bandwidth depends on the services to be provided and how the major transmitting stations on the ground are designed to deal with the demand.

As we have emphasized, bandwidth is our most valuable commodity; it is expensive. Just think what it has cost to put 800MHz into space! So it is essential that we use bandwidth most effectively and for services that provide the highest return. We do this by a number of very ingenious means including sharing the bandwidth among several users in different parts of the globe at the same time (their geographical separation and the focusing of the antennae avoids interference), or we assign bandwidth to different beams so that in one place the bandwidth is used for the global beam but in another it is used for a spot beam. Frequencies are also reused by polarizing the signals differently, in this case receiving and transmitting two different messages using the same frequencies but one polarized horizontally and the other vertically, just as you would polarize light and transmit two different pictures on the same light beam. We shall have more to to say about frequency reuse when we explore the use of digital communications.

The transponder is a repeater, or a relay. Therefore, it must respond to the instructions it receives from the earth transmitting station. Let's now examine these instructions.

THE EARTH SEGMENT

For every satellite in orbit there may be several hundred, if not thousands, of earth stations. There are considerably more than 10,000 earth stations in the United States alone receiving transmissions from domestic satellites serving the cable television and broadcast industries. Because the footprints of the domestic satellites are much smaller than those served by INTELSAT, these earth stations are likely to be smaller and less expensive than those of INTELSAT which provides global or hemispheric coverage for its customers in 109 or more countries. The point here is that the investment on the ground is much, much greater than that in space. This is typical of broadcast systems that generate signals from a single point to many locations. For example, it has been estimated that for every dollar spent for the capital equipment to build a television station, the public has spent about $10. It might be interesting to calculate the radio for satellite systems.

There are two types of earth stations—those that both transmit and receive signals and those that only receive. The former, of which there are relatively few, are the very large and expensive standard transmit-receive earth stations in the INTELSAT system, and the main control stations that serve the domestic broadcast satellites and the emerging business satellites such as Satellite Business Systems (SBS). The latter are called TVRO (television receive-only earth stations). We begin our discussion of the earth station with the main control station, for this is where the services a satellite is designed to deliver begin.

The Transmitting Earth Station

The extraordinary flexibility of modern telecommunications—the ability of today's communications systems to deliver the wide range of services demanded of them—is best illustrated in the functions of the earth transmitting station. Here, too, the marriage of communications and computing is shown in all its glory.

An earth station has three basic components—the very visible antenna system, the amplifier/converter often referred to as the "front end," and the transmitter/modulator for the transmit station, and, its counterpart, the receiver-demodulator for the receive-only stations.

Signals to be transmitted—computer data pulses, telephone voice conversations, video, or facsimile—are delivered to the transmitter from various originating points. It is here that the signals are processed to accommodate as many uses and users as possible. Once processed, the signals are delivered to the high-powered amplifier (the TWT described previously) for transmission to the satellite repeater or transponder. At the start of the space trip, while still in the earth's atmosphere, there is likely to be a great deal of electromagnetic noise interference. But once in outer space and out of the earth's atmosphere, there is much less interference. The signal's trip back to earth is a long one, perhaps as much as 1000 times longer than usually traversed between ground-based antennae. The strength of the signal radiated by the antenna diminishes as the square of the distance traveled, so what actually reaches the spacecraft will be very small, in the neighborhood of one million millionth of a watt. The same degree of loss occurs after the signal is amplified by the transponder and sent back to earth. Add to these losses the unpredictable interference that occurs due to bad weather. For the higher frequency systems, these weather losses are even more severe. Here lies another opportunity for a design tradeoff. A choice must be made between the gain or amplification the transmitter must have, the gain the transponder on board the vehicle must provide, and the amplification or gain the receiving station will have to add in order for intelligent communications to take place.

On satellite footprint maps you will often see the term EIRP, or effective isotropic radiated power. EIRP is a measure of the power level of the signal leaving the satellite antenna and of the power level of the signal leaving the antenna at the earth station. There is a tradeoff between the amount of power in the transmitting earth station and the amount of power in the satellite. Clearly, the more power the transmitter has the less that might be needed in the orbiting vehicle. But if there is less power in the vehicle, larger receiving antennae would be required. Since there are fewer major transmitting stations than there are receiving stations, it makes sense to keep the cost of the receiving stations down. Hence, much more power is put into the transmitting systems, both in the vehicle and on the ground.

We have said it many times and will repeat it once again: bandwidth costs money. Anything we can do to make the most efficient use of bandwidth results in more cost-efficient use of communications and lower costs to users.

As you might expect (see Chapter 2), FM is the preferred mode of signal modulation for satellite communications. You will recall that FM offers the best chance for obtaining adequate signal-to-noise ratios, and satellites often operate in electromagnetically noisy environments. To date, satellites have used frequency division multiplexing, and as a result, you will often come across the letters FM/FDMA which stand for frequency modulation/frequency division multiple access operation.

Recalling our discussions about bandwidth in Chapter 2, we found that the bandwidth for television is about 6MHz or 6,000,000 cycles per second and that bandwidth for voice is about 4KHz or 4000 cycles per second. You will recall, too, that the capacity of a transmission channel is determined by four parameters: time, bandwidth, signal strength, power, and noise. Let's examine this relationship a little more carefully.

We know that power is especially critical in the design of satellites. The long distances to and from space means that receiving amplifiers pick up extremely small signals, and we can only create so much power in the satellite in order to send more powerful ones. Consequently, the tradeoff between bandwidth and channel capacity is critical to the effective utilization of a satellite transponder.

Channel capacity is directly proportional to its bandwidth and signal-to-noise ratio (S/N), or to put it another way, bandwidth is inversely related to noise. In other words, if the noise in the channel is large, more bandwidth will be needed to transmit the signal without losing information.

Of course we want to utilize the valuable space in the transponder as efficiently as possible. To do so we might wish to send not one but three 6MHz color television signals, or one television signal and a great deal of voice and data. However, this would require an enormous amount of power in order to be assured that intelligent signals are to be received. We do not have that power, at least not yet, so we make a tradeoff and use an entire transponder, for the high-valued television programming, and use other transponders for voice and data.

Because there are different market values for different varieties of information, satellites mix their media transmissions. For example, the WESTAR satellite provides one color television channel, 1200 voice channels, sixteen channels for data transmission at the rate of 1.544MBs, 400 channels for data transmission at 64,000 bits per second (kilobits per second or KBs), and 600 channels for data transmission at 40KBs, with some room left over. These uses take place simultaneously, and many users access the satellite system, hence the term "multiple access." The first INTELSAT satellite launched in 1971 provided 4000 voice circuits and two color television channels; INTELSAT V, launched in 1981, has a capacity to provide 12,000 two-way voice circuits plus two color television channels. Of course, other combinations are possible, including simultaneous access for voice, data and television. Facsimile also can be added to the picture. Because of the clever ways beams can be separated, INTELSAT V also could be used to send up to eighty TV channels around the world, if nations would agree to the interchange.

If a 36MHz transponder can send a 6MHz color television signal, how many voice or data signals could that transponder transmit? Let us say that a television channel of 6MHz can handle about 100,000,000 bits per second or 100 megabits per second (MBs). This means that our 36MHz transponder also could be used to transmit and receive about 300MBs. If all of the transponders were to transmit data, the satellite would be able to transmit the entire *Encyclopaedia Britannica* some six times from one continent to another in one minute flat. Geewhiz exercises like this are quite popular in the satellite business.

As noted in Chapter 2, there is much to be gained from digital communications, and had we had the chip and a few other technologies when the telephone was invented, we might very well be entirely digital today. The most important benefit of digital communications is that it allows for time division multiplexing; thus more can go into the bandwidth than would be possible with frequency division multiplexing. Satellites are going digital and very rapidly. Time division multiplexing, as described in Chapter 2, is one means to increase their capacity. Indeed, INTELSAT VI will be a time division multiple access (TDMA) satellite with the capability to more than triple the telephone capacity of INTELSAT V.

In TDMA, a given bandwidth channel in the up- and downlinks is shared by a number of digital transmissions from different earth stations. The transmissions are coordinated by the earth stations so that they are synchronized with respect to each other and appear sequentially at the input to the satellite, much as we showed the several well-ordered signals in Figure 2-19, Chapter 2. A switching capability is added in the satellite that permits the interconnection between different up- and downlinks to be dynamically changed in synchronism with the signals from the respective earth stations, thereby keeping track of which portion of the signal belongs to which earth station and to which earth station it is to be delivered (see Figure 4-13). Note that signal (a) from Station 1 (ST-1) is delivered to Station 5 (ST-5) and that signal (d) from Station 4 (ST-4) is delivered to Station 2 (ST-2).

By adding the ability to divide up the area to be covered into geographical zones by means of spot beams, we can significantly increase the *connectivity* of the satellite—that is, the ability to connect more terminals to the satellite, to expand the satellite network, and provide services to more people. In short, we are increasing the utilization of the satellite bandwidth which helps to lower the cost for each user of a portion of that bandwidth. (We shall have more to say about connectivity and networks in the next chapter.)

The number of transmitting earth stations is increasing rapidly, not only in the U.S. but throughout the world. The variety of earth stations is also increasing; there are mobile stations; transmit facilities for business offering voice and data primarily with occasional video; transmitting earth stations for broadcasting; and the main control stations for INTELSAT that provide voice, video, data, and image transmissions throughout the world. These facilities have become smaller, cheaper and, as we have noted, even mobile.

What is particularly important about these developments is that the satellite is becoming an ever more flexible instrument for communications, capable of

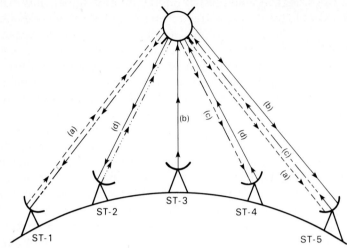

FIGURE 4-13
Time Division Multiple Access.

offering many varieties of services and at costs that are falling. We shall explore these services later, but it might be well to note that the *space segment* cost per circuit (that is, for two-way voice or telephone service) has fallen from about $33,000 per year in 1965 to about $850/year in 1980; with TDMA, this cost can be expected to be in the neighborhood of $300 per year. Remember, however, this is only for the space segment! There is still the ground segment and the troublesome "final mile" to be discussed.

The Television Receive-Only Earth Station

TVROs are more numerous and are what we predominantly see in our travels in cities and in the countryside. They, too, are composed of three major components: the antenna, the "front end," and the receiver/demodulator.

The antenna system is pointed to a particular satellite. The better the aim, the more energy the antenna can pick up, and, consequently, the greater its ability to receive the very weak signal coming from the satellite. Because the global beams emanating from INTELSAT are so large in area, they deliver a small signal and a large "dish," as much as ten feet in diameter or more than seventy-five square feet in area, is required. As the beams are more focused, the power delivered, as indicated on footprint maps as EIRP, is higher and the area of the TVRO antenna is smaller. EIRP, to refresh your memory, is the measure of the power level of the signal either leaving the satellite antenna or the transmitting earth station antenna.

The signal is sent by the antenna into the low-noise amplifier which takes this very weak signal and amplifies it with as little noise added as possible. As noted in Chapter 2, when you amplify any signal that already has noise you also amplify the noise, and most electronic hardware generates noise on its own. This condition could be disastrous if the incoming signal is already very weak and somewhat noisy. Consequently, a whole new breed of amplifiers was developed for satellite receivers; the low-noise amplifier or LNA is at the heart of earth station design and critical to the success of satellite communications.

The final element in the chain is transforming the signal from the satellite into audio, video, or digital pulses. This transformation is performed by the receiver.

The efficient utilization of bandwidth begins in the transmitting station and is paralleled in the receiving station. The earth stations must reflect the user's needs as these needs define the functional requirements for the space vehicle. The communications and computer services required by the clients of satellite communications define how the bandwidth in a transponder and, indeed, how the entire bandwidth space in the vehicle, is to be allocated.

For example, the Satcom and Galaxy families of satellites are devoted almost entirely to the delivery of television programming to cable systems and to other television broadcasters. Several thousand TVROs are located at cable system headends and at broadcast stations, while the relatively few major control centers are at locations on the east and west coasts of the United States. The IBM/Aetna/Comsat joint venture that launched the SBS series of satellites (Satellites for Business Services), provided a mix of services including private-line telephone, high-speed data transmission facsimile, the transmission of still pictures, and at least two forms of video for teleconferencing, slow-scan and full-motion video. The SBS system is a fully digital system, all of its transmissions are digital, using very sophisticated forms of time division multiplexing requiring complex scheduling procedures to ensure that the right messages reach the right people. SBS earth stations are usually mounted at the firms subscribing to SBS services, often on the roof of a headquarters building. What happens to the long distance and local telephone carriers in this process? We shall look into this solution to the "final mile" a bit later, but these are the sort of technological developments that have caused the recent reshuffling of the nation's telecommunications infrastructure, including the dismemberment of the staid old Bell System.

The earth station is not an independent component, rather it is but one component in a system and its design and selection is dependent on the many factors that constitute the system. Remember the Shannon-Weaver Model of Figure 4-6? The earth station is the receiver in the complex communication system whose design depends on the transmitter, the transmission medium, and the noise in the system.

Compare the earth station to your high fidelity system. In both cases you are receiving or picking up a very weak signal and you want to amplify that signal in a manner that assures very high fidelity. The satellite signal is probably much smaller and weaker than the signal the stylus delivers to your hi-fi amplifier. Moreover, the stylus is pretty much locked into a groove, but the satellite may be wobbling in space 22,300 miles away, other satellites may be relatively nearby, the earth is moving, and you must aim your TVRO antenna as accurately as possible at the "bird" you want to "see."

The problem of noise is a concern not only in space but perhaps even more so on the ground. In a major city there are few locations free from the electromagnetic noise of other radio signals: broadcasters, mobile radios, pagers, automobile ignition systems, etc. It often seems that the location best suited to the plant engineer, campus architect, or city planner is in the direct beam of someone's transmitter. That is why you often find TVROs mounted in the "shade" of large buildings or even below ground level. These issues are even more troublesome for major control stations. You do not want to begin the transmitting sequence with noise from your surroundings and, furthermore, you cannot interfere with your neighbors' radio communications.

In our discussion of the satellite we found that the cost of the space segment is falling rapidly. In a period of less than twenty years, the cost per voice channel per year fell from about $16,000 (in 1965) to $330 (in 1964) or, in 1984 dollars, from $57,000 to $370. Also, during this period, the lifetimes of the satellite in orbit rose from a mere 1.5 to seven years. Are similar cost reductions likely to occur on the ground, where they would really have an impact on the users?

This depends on what the users want and how many there are. And the number of users depends, in turn, to some considerable extent on the ease with which they can access the system—in other words, the "final mile."

THE FINAL MILE

The telephone system can be accessed merely by picking up the instrument, waiting for the dial tone, and dialing the call. Radio or television programs are accessed by turning on receivers. It is more difficult to access the satellite system, but it is becoming somewhat easier. Accessing satellite communications requires us to traverse what has been called the "final mile." After a 44,600-mile journey to and from space, messages must get to and from the earth station in order to be transmitted and received.

Earth stations are not telephones, nor are they radio or television sets. Aside from cost, there are many reasons for not having one next door to your office or home—interference and noise problems, difficulty in aiming at the appropriate vehicle, and the growing number of vehicles at which to aim.

The technologies of the final mile are almost as numerous as the uses to which you might want to put satellite communications. These technologies include both wired and broadcast technologies and much of this book is devoted to discussing them. They are listed in Table 4-1.

Through its long-distance operations, the telephone company provides both the master control station and the TVRO for your telephone services. The existing telephone system, referred to in Chapter 6 as the local loop, provides the final mile to your homes and offices. As we shall see, the telephone system is very flexible with the ability to deliver voice, data, and a form of video

TABLE 4-1
THE MANY FINAL MILES

From satellite to	To	Via
Local Telephone Carrier	Homes and Offices	Twisted Pair
Local Broadcaster	Homes	Broadcast Transmission
Cable Television Headend	Homes & Office	Cable
Multipoint Distribution System	Homes & Office	Microwave
Satellite Master Antenna	Apartments & Condominiums	Cable
Instructional TV Fixed Services Systems	Schools	Microwave
Cellular Radio Network	Automobile, Airplane, Train	Microwave
Homes and Offices		DBS

referred to as *slow-scan*. Motion in slow-scan video is, in a way, captured by a series of snapshots which can be transmitted by the relatively narrow bandwidth telephone system.

Cable television companies receive their local and network broadcast signals at a "headend" and then send them along the cable to your home. Their headend sites also have several TVRO stations for receiving the pay-TV as well as superstation programs and an increasing number of services delivered nationally to all cable operators who purchase them for retransmission and resale to their subscribers. Satellite antenna "farms" are rapidly becoming common sites on the landscape. Domestic satellites, such as the previously mentioned Satcom and Galaxy series, are television delivery satellites and use cable for their final mile delivery to your television receiver.

Other domestic satellites such as WESTAR which, as noted earlier, provide a mix of services including video conferencing for business and educational institutions, often by-pass the final mile entirely. A mobile TVRO can be provided on site for the duration of a teleconference. To send as well as receive video, in a conference known as "two-way video and voice" or "synchronous teleconferencing," you would access the transmitting station either by telephone, cable, or one of the several broadcast modes listed in Table 4-1.

THE DIRECT BROADCAST SATELLITE

In the chapter on broadcasting, we will see that many new local and community broadcast technologies are emerging that can serve as final miles for satellite access. They are made possible by improvements in frequency multiplexing. We have learned to use the broadcast spectrum more efficiently by cramming more signals into the limited space. We also have opened up new spaces or higher frequencies just as satellites are doing. The technology for local communication services has developed almost as rapidly as have the technologies for long-distance services; the satellite is but one part of the communications technology revolution. There are instructional television fixed services (ITFS) systems for reaching from satellites into schools, multipoint distribution services (MDS) for connecting homes to entertainment services and businesses to data services, and mobile radio networks for interconnecting people on the move to satellites. There also is the direct broadcast satellite (DBS) which may do away with the need for the final mile.

With the ability to focus a satellite beam ever more narrowly with more power, will it be possible to deliver a signal to a small antenna mounted on the roofs of houses or businesses without having to go through any final miles? The answer depends on several factors, not the least of which is determining your intended use for satellite transmission.

We have seen that the amount of bandwidth required for a video is about a thousand times greater than that required for a voice signal and that a voice signal requires, roughly, about a thousand times greater bandwidth than that required for data transmission. Some engineers have sought to relate this difference in bandwidth requirement to the complexity and cost of hardware and it is useful to consider this although the numbers are not terribly exact. It could be argued, however, that earth stations that transmit and receive data only could be smaller and less expensive than those that transmit and receive voice and very much smaller and less expensive than those that transmit and receive television. So if you wish to use a satellite for data only, in the very near future you will not have to traverse the final mile; a small earth station on your roof or in your parking lot at a cost of less than $1000 will do. What is more likely to happen is that several satellite users in a single building will share the bandwidth of a data- or voice-only satellite and could readily afford a TVRO on their building rooftop, thereby by-passing the final mile. This is already occurring as businesses mount even larger antennae on the roofs of their office buildings and factories take advantage of satellite services directly rather than through existing earth-based final miles. Some movie and television buffs are already installing their own TVROs in backyards. People in remote locations not likely to be wired by cable are installing TVROs of their own. Recognizing the importance of television to the outback, the Australian Broadcasting Company hopes to provide a system that will enable farmers to receive programs with "dishes" costing no more than $1000.

Direct broadcast satellites (DBS) will be able to deliver signals directly to antennae approximately one foot in diameter. If developments in amplifier design continue as they have in the recent past, there is reason to expect that an entire TVRO can be mounted on a roof or on a television receiver. The goal of Japanese and U.S. industry is to make such TVROs available for as little as $300, perhaps even $100.

Needless to say, these developments will drastically alter the nature of the nation's telecommunications and broadcast infrastructures. What happens to the local telephone company? What about local broadcasting? To say the least, these are issues of great concern to Congress, the FCC, and to businesses that are delivering local communications.

THE USES TO WHICH SATELLITES CAN BE PUT

Developments in satellite technology neatly reflect the ability of the chip to bring computers together with communications, for the semiconductor chip is directly responsible for helping communications engineers make such efficient use of bandwidth. The economic attractions of the satellite rest not only on its ability to link distant parts of the earth with a great deal less difficulty and, of course, lower cost than is possible with earth-based technology, but also on the increased amount of effective bandwidth that can be obtained from the satellite transponders through frequency reuse, digital transmission, and TDMA. We have seen how the cost of circuits on satellites is falling rapidly each year and, with the space shuttle placing ever larger satellites in orbit, we can expect these costs to continue to fall.

With lower transmission costs have come increasing demands for service, and, in turn, a decline in the cost of the earth stations themselves. It used to be that a TVRO for a cable television system or broadcaster cost in the neighborhood of $150,000. In 1982, the average purchase price of a TVRO for a 4/6GHz operation was about $15,000. Television junkies are now able to buy TVROs for home use for about $3500, fully assembled, and DBS receivers are advertised for sale at no more than $300 installed, in large quantities.

The satellite liberates us from the tyranny of space and time and is rapidly liberating us from bandwidth limitations. Satellites already play important roles in today's communications and their use will continue to expand. The best way to determine just what the satellite can do for you is, simply, to look at what communications services you use today and how much you spend for them. This is not an easy task. That we have not entirely entered the information or telecommunication era is evidenced by how difficult it is to determine how much the nation spends for different types of communications services. The best we can do is accept and try to draw some conclusions from the data available. Let's put some bits and pieces together to get a picture of our use of telecommunications today.

In 1981 there were 47 million business telephones in use and 135 million residence telephones in 97 percent of all households. Together, these telephones were used to send more than 900 million messages. Operating income from the domestic use of telephones was greater than $55 billion with another $1.5 billion derived from international services. About 30 percent of the telephone traffic was long-distance, the kind likely to be delivered by satellite.

Of course, the telegraph has not been forgotten; telegraph traffic, including Telex and data, accounted for about 5 percent of telecommunications applications in 1981. More than $256 million was spent on domestic telegraph services and $380 million on overseas services.

Telephone traffic is expected to grow at about 8 percent per year, Telex and data traffic at about 15 percent per year. With the growth of transnational businesses and facilities expansion by domestic firms throughout the United States, we might expect that a healthy proportion of the rise in telecommunications traffic will be for long-distance services, the most likely candidates for satellite delivery.

To this overview of message traffic, we should add what little we can deduce about broadcast and cable communications from the Statistical Abstracts of the United States. With more than 38,000 radio and television broadcast stations operating in 1981, a mixed bag of commercial, public, private, educational, translators, or repeaters that deliver signals to locations out of range of the broadcast station, we know that a great deal of long-distance transmission of voice, video, and, increasingly, data traffic is carried throughout the land. In addition, another 30,000 common carrier broadcast facilities, selling point-to-point microwave transmission services, are in operation. Just what these facilities deliver—how much voice, video, data, or facsimile—is not as yet recorded. But we do know that the satellite is a major addition to these modes of information delivery, especially where the distance between transmitter and receiver is more than 1000 miles.

Projecting from the best available data and being quite conservative about whether or not satellites will be most efficient for long-distance services (greater than 1000 miles) only, we might project the uses of satellites in 1990 as in Figure 4-14.

Let's be a bit more adventurous and consider what might happen as satellites make more efficient use of spot beams and the TDMA mode of operation.

FIGURE 4-14
How We May Be Using Satellites in 1990.

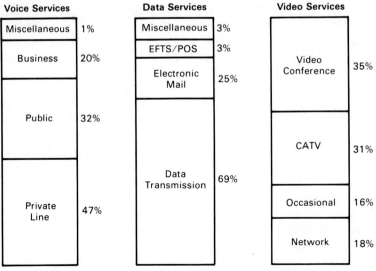

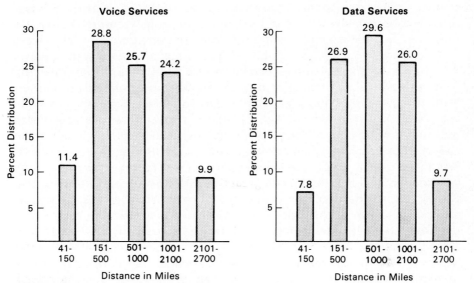

FIGURE 4-15
Traffic Demand Versus Distance for Satellites in 1990.

Satellite delivery would become effective over smaller distances, perhaps down to well under 100 miles. Indeed, proposals have been made for "street-corner-to-street-corner" satellite-delivered signals.

Now let's look at how the demand for services, data, and voice, varies with respect to distance. This is illustrated in Figure 4-15. Even if we assume that the satellite will only be cost efficient for distances of greater than 1000 miles, it looks as if more than 70 percent of our traffic could be via satellite. With the way satellite technology is now going, however, it is certain that the cost efficiencies will be available even over distances as short as 200 miles. Does this mean we might use satellites to talk from city to suburbs, or even within some of our larger cities, such as Los Angeles and Houston?

THE FUTURE

Developments in telecommunications and computer technologies, or what we call the information technologies, are creating competitors for the satellite—indeed, creating competition for almost every new development they spawn. Consider, for example, the optical fiber cable. In Chapter 2 we noted its ability to deliver unlimited bandwidth and essentially noiseless transmission. These cables are finding their places not only as replacements for land-based microwaves but, perhaps even more important, as part of the new generation of transoceanic cables now being laid and planned.

Cables are attractive investments since they return revenue directly to the nations that install them. The INTELSAT system as an international consortium, returns its investment to the 109 and more participating nations as a percentage of each country's investment and in proportion to its use of INTELSAT. The United States, for example, is a 23 percent shareholder in INTELSAT and receives a percentage of its investment in accordance with use. Indonesia is a 2.4 percent shareholder and receives its return based on a proportion of that ownership and its use of the system. At the present time, Indonesia's communications traffic is probably greater among its Southeast Asian neighbors than with other nations of the world. Clearly, if Indonesia wishes to generate revenues from communications, it might do well with its own regional satellite and with the installation of undersea cable between itself and other nations. In fact, that is exactly the course Indonesia has chosen to follow. As noted at the beginning of this chapter, Indonesia has launched a multiple satellite system, Palapa, to provide domestic communications for itself and its neighbors, especially those in ASEAN, and is a partner in one or more transoceanic cable operations between the Philippines, Hong Kong, and Canada. There is every reason to believe that these cables will be optical fiber cables, with enormous bandwidth capable of providing transmission services at very low cost. Likewise, there is every reason to expect that other nations will follow the same course.

The human need for communications seems to be unlimited. Throughout the history of communications, new technologies have not replaced older technologies; radio did not do away with the newspaper, and television did not destroy radio. So optical transmission systems will take their place in the spectrum of transmission technologies as the satellite has done. In the future we can expect that the cost of transmission will continue to fall as the competition among transmission technologies heats up and that is all to the good.

There may, however, be limits to satellite growth as the geosynchronous orbit becomes filled. New frequencies, along with the introduction of digital transmission and of techniques for compressing digital signals, thereby enabling even more efficient use of bandwidth, will delay this limitation on orbital spaces for satellites for some years to come. What will not be delayed are the international policy issues created by modern telecommunications and especially the satellite.

It has become increasingly clear over the past decade or so that the separation between national and international communications has blurred. Furthermore, the separation between national and international business also has become much less distinct. Every major bank wishing to operate internationally must be a member of the international SWIFT network (Society for Worldwide Interbank Financial Telecommunications) and with few exceptions, national airlines that wish to operate in the international sphere must join the SITA system (Societe Internationale de Telecommunications Aeronautiques). These are examples of international businesses requiring communications across national boundaries. Domestic communications policy considerations are, consequently, very much linked to international communications policy considerations. Given the desire of nations to maintain their sovereignty as well as their opportunities for favorable positions on the international trade routes, future negotiations concerning orbital slots, assignment of frequencies and standards for computer communications are certain to be complex and difficult. The future is bright with promise and exciting for those who thrive on problem solving. To paraphrase the familiar Chinese curse, "we live in interesting times."

ADDITIONAL READINGS

Surprisingly, there are relatively few works dealing seriously, yet not technically, with the satellite. Much has been written for weekly magazines and newspapers, generally extolling the virtues of space communications without regard for competitive systems and the social issues that could emerge. Another body of literature has dealt with the international politics of satellites. We have selected a mix of sources that may satisfy your desire for additional information about all of these topics.

Brown, Martin P. Jr., ed, *Compendium of Communication and Broadcast Satellites, 1958–1980* (New York: IEEE Press, 1981). The number of communication and broadcast satellites is proliferating so fast it is often difficult to keep track of them.

Clarke, Arthur C., *Voices from the Sky* (New York: Pocket Books, 1980). The father of the communications satellite tells his story.

Glatzer, Hal, *The Birds of Babel: Satellites for the Human World* (Indianapolis: Howard W. Sams, Inc., 1983). A "quick read" that offers a glimpse into the world of satellites.

Martin, James, *Communications Satellite Systems* (Englewood Cliffs, NJ: Prentice-Hall, 1978). A comprehensive text that is an excellent reference for those who want to know more.

Miya, K., *Satellite Communications Technology* (Tokyo: KDD Engineering and Consulting, Inc., 1981). For the engineer or the very curious.

Pelton, Joseph W., *Global Talk* (Rockville, MD: Sijthoff & Noordhoff, 1981). A superb prize-winning book by one who knows the business from the beginning.

CHAPTER 5

NETWORKS AND TERMINALS: THE ARCHITECTURE OF TELECOMMUNICATIONS

We first encountered the idea of a network in Chapter 2, and we saw how important and useful this idea is to the understanding of modern telecommunications. In this chapter, we examine networks in more detail. How networks and the terminals they interconnect create a network architecture and how that determines what you can do with telecommunications systems are our subjects.

The term network has suddenly become fashionable. We don't talk to people, we "network" with them. We are no longer simply members of a club or group, we are "a network." We have adopted the language of communications engineering in our daily lives because networking implies a rich and flexible linkage of communications possibilities, exactly the quality we desire in our social relationships.

Networking is what modern telecommunications is all about. In this chapter we discuss the idea of a network and show why this important concept has become fundamental to modern telecommunications. In so doing we encounter the ever-present terminal, that necessary link into and out of telecommunications networks.

THE ARCHITECTURE OF NETWORKS

Information is valuable only if it is used—communicated and transformed into knowledge and actions. Networks provide the pathways to achieve this, and in that regard, an information society is a society of information networks.

These networks have a pattern or structure often especially suited to some particular use. Computer and communications people have labeled this pattern the *network architecture*. When engineers discuss the architecture of a network they are describing how the physical parts of the network—the terminals and switches—are organized into a highway for the delivery of information. This architecture defines how information flows are managed on that network and how and when information is presented, interpreted, and transformed while traveling along that highway.

In Chapter 2 we showed why a network is so important for communications. Networks provide the connectivity we desire and, if properly designed, they can do so quite efficiently, making the most of valuable bandwidth while conserving copper (see Figure 2-2). Perhaps the most familiar network architecture is the telephone system. Figure 5-1 illustrates this structure in its simplest form. The terminals shown here all perform the same function. They are telephones, transmitting and receiving human communications. This is the telephone architecture you find in your calling zone or neighborhood. In telephone terminology it is the fifth level or class 5 switching system (counting from the highest level down). This type of structure is often said to have a "star topology." Every terminal in the star can be connected to every other terminal through the switch; hence we can say that in this structure there is complete connectivity.

As you leave your neighborhood, the architecture becomes somewhat more complex (see Figure 5-2). Note that in this illustration there are several star topologies—that is, several nodes or switching centers with many terminals of telephones. They might be neighboring telephone areas or zones or long-distance networks which we will discuss in greater depth in Chapter 6. A convenient way of looking at these different nodes is to see them as sitting on several levels on the network. This leads to the concept of a network hierarchy.

In large telephone systems, a switch in a major city would be the node or terminal point into which many neighborhood star networks feed and then are connected to other neighborhood switches or nodes. The switching-center node in the major city is said to be higher in the hierarchy than those at the

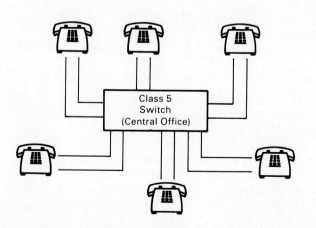

FIGURE 5-1
Star Topology Has Complete Connectivity Through the Switch.

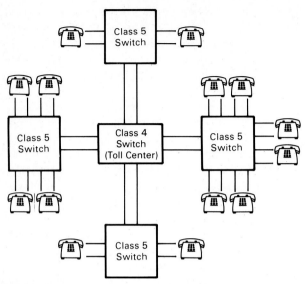

FIGURE 5-2
A Two-Level Hierarchy.

neighborhood level. The levels within the network where different communication tasks are performed describe the network hierarchy. How many levels a communication system needs is determined by what you expect information to do and how you intend to manage it. The hierarchical structures of networks also determine, to a considerable degree, the kind of technology used for communications; the more functions performed at a level, the more likely that a sophisticated (and often more expensive) technology will be used. The satellite described in the previous chapter certainly performs significant roles in the world's communication networks and might very well be classified at the highest level—no pun intended. The cost of a satellite therefore matches its place in the hierarchy of communication architectures. In the next chapter we shall see how the telephone system uses network hierarchies to delineate local loop communications from long-distance communications and their respective technologies.

Cable television systems have yet another form of network topology as illustrated in Figure 5-3. In this configuration, we begin at the headend, or control center, of the cable system where signals originate or are prepared for retransmission after having been received from satellites and over-the-air broadcast stations. These signals form several large branches, or trunks, and then smaller branches which deliver cable television services to subscribers. This is, logically enough, called a *tree structure*. If the system becomes too large to be served by a single headend (a condition that occurs when the trunks of the trees are longer than about ten miles), smaller headends are established. These are the hubs (A, B, and C) shown in Figure 5-4. The tree

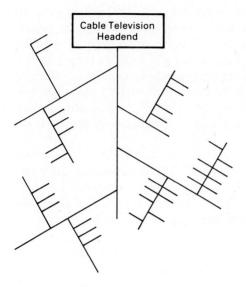

FIGURE 5-3
A Tree Topology Used in Cable Television Systems.

structure of Figure 5-3 is now a multilevel structure, in this particular case, a two-level system, with the headend at the top of the hierarchy and the hubs at the second level. The hubs do not perform all of the functions of the headend—for example, signals do not originate at the hubs as they can at the headend. Consequently, the complexity and cost of the hub technology are significantly less than that of the headend.

FIGURE 5-4
Two-Level Multihub Structure.

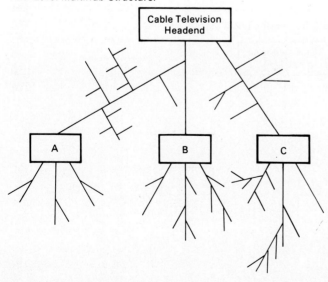

Let us suppose, however, that the community served by hub A is an Hispanic-speaking community and that you, a responsible cable programmer, wish to serve its special needs. These may include needs not required elsewhere such as films, news, and public affairs programming in Spanish. Hub A can be upgraded by adding both programming and switching capability, perhaps even a star switch to form a local star network as illustrated in Figure 5-5. Another level has been created in the hierarchy through the addition of more complex technology needed to perform special functions on the network serving that community.

Network topologies describe configurations of terminals, switches, and other hardware in the communication system, while network hierarchies describe the functions of the system. In general you will probably be more interested in hierarchies than in topologies, leaving to the designers and engineers the task of selecting the best topology to serve your functional needs.

The existence of physical connections—cable, telephone, or satellite-delivered transmission—between the entities on the network does not guarantee communication between these entities. As in face-to-face conversation, certain rules govern communication, such as your relationship with the person to whom you are speaking, his or her social and professional status, your feelings about your conversational partner and how you think that partner feels about you, and where the conversation is taking place. All these factors play important roles in ensuring that the conversation is meaningful and, hopefully, intelligent. Likewise, certain rules are required to ensure that participants on a network can communicate with each other. These rules are called *protocols*. Protocols are the standards networks require to ensure that all participants can communicate intelligently and meaningfully with each other.

FIGURE 5-5
Multilevel Hierarchical Structure: Tree and Local Star.

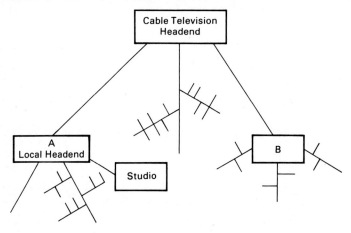

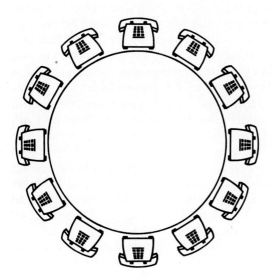

FIGURE 5-6
A Ring Network.

Protocols are procedures for the exchange of information. Telephone protocols have been standardized for more than 100 years and are embedded in the technology in what may be called telephone etiquette. But while we need no longer be too concerned with the protocols for voice communication on the telephone, we must be concerned about the rules of networks for video and data communications.

In later chapters it will become evident that the concepts of network topology and hierarchy are very important in your choice of information system and, often, to the politics of information. Consider, for example, what could happen if you were using the network shown in Figure 5-6 with all of the telephones arranged around a ring (called a ring network). Connectivity between every member of the network is performed in sequence; every message from any terminal on that ring could be received by everyone else and conceivably the party nearest to the source could hear the message first. Many international news networks operate this way; the country furthest away from the source may feel a bit left out by the time the news reaches its borders. You do not need much imagination to recognize that a great deal of interesting political theories surround news network architectures of that kind.

SWITCHED AND NON-SWITCHED NETWORKS

Networks may be switched or non-switched. The telephone network is a switched network; it allows any user to be connected to any other user. Broadcast systems are non-switched systems; the same message goes to everyone who is tuned in. Cable television systems are non-switched systems; all of the channels are delivered to all of the subscribers and the subscribers choose what they want to watch using a terminal (the converter) in their homes.

The telephone network is a *line- or circuit-switched* system; when you place a call, the telephone switching system locates a circuit or a line for your call before it is put through and you stay connected on that line as long as you continue to talk. This type of system is illustrated in Figure 5-7(a).

When telegraph systems spread throughout the nation and the world, messages often had to travel on several different networks in order to get from one point to another. Telegraph operators at intermediate points would receive the message on one network and transfer it to another network. Many times the intermediate office would have several networks coming in and going out and the operator would transfer, or switch, messages between networks as shown in Figure 5-7(b). Today, this is performed by automatic switches in *message-switched* systems.

Because it is the actual message that is switched rather than a transmission technology as in line- or circuit-switched systems, the message can be held or

FIGURE 5-7
Switching Styles.

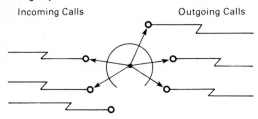
(a) Line/Current Switching

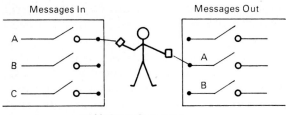
(b) Message Switching

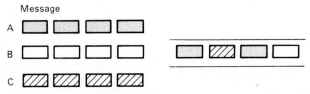
(c) Packet Switching

stored before sending forward. Airline reservation systems do this in order to deal with time zone differences between a travel agent's office and their reservation computers. Message store and forward is also useful when reservations overload offices or computers; they can be lined up and dealt with in order of their arrival. You encounter the storing and forwarding of messages every day—when you speak to an answering machine on the other end of your telephone, for example. It, too, is an application of message store and forward. The old telegrapher's operating style is now being incorporated into modern computer-communications systems as electronic mail, a subject about which we will learn more in the next to last section of this book on *network information services*.

When telephone and other traffic—*message traffic* in the lingo of the engineers—gets heavy, there may not be enough lines to send the messages let alone store them for later delivery. As we discovered in Chapter 2, there is a great deal of empty space in any message signal, whether it is a voice or data signal. As you will recall, these empty spaces can be used for time division multiplexing (see Chapter 2). If it were possible to break up messages into convenient chunks, identify these chunks so that we know which piece fits with which piece, we could intersperse different messages using all of the time spaces or slots along whatever circuits or lines the switch finds for us. This type of switching is called *packet switching,* so named because the chunks are known as packets. An example of packet switching is shown in Figure 5-7(c). We shall have more to say about it in our discussion of computers in Part 5.

TERMINALS: INTELLIGENT AND NOT-SO-INTELLIGENT

For communication to occur, there must be a point at which information can enter and leave a network. However, there is no need for a live actor as long as terminals are present. What are these devices that do not depend on a human and yet are required for the transmission and reception of information?

A terminal is any device capable of sending and/or receiving information over a communications channel. It is a device for entering your voice on a communications channel, and in this sense, your telephone is a terminal. It is a device for listening to music—your radio high-fidelity set, for example—or, for looking at video images—your television set. A terminal is a device for entering data into computer systems and it is the means by which the decisions computer systems make are communicated to the environment these decisions are meant to influence. A terminal can be the thermostat in your home which communicates desired temperature to your heating and cooling system as well as the device that tells you and the temperature controller just what the temperature actually is.

The telegraph key was perhaps our first terminal for electrical communications. The keyboard on a personal computer is a terminal for the entry of words and data to be processed and sometimes to be communicated to other terminals that deliver information to other computers. The CRT (cathode ray tube) on which words and data are displayed is also a terminal.

The device on which temperature differences are measured by changing conductivity and which communicates this information (your thermostat) is a kind of semiconductor known as a thermocouple. It is also a terminal as are pressure gauges, thermometers, and other instruments for detecting and displaying information.

Terminals can be classified as input devices, as in the case of equipment for delivering information to the networks for communication such as the keyboards of computers and word processors or the turntable for your hi-fi system. Terminals also can be classified as output devices, such as printers, the CRT on your computer or television set, the speakers in your hi-fi system, and light displays in automobile dashboards and on microwave ovens.

Many terminals perform only one task and are therefore considered "dumb." Not too many of these devices exist today, even in consumer goods, since the semiconductor chip makes the addition of memory and information processing relatively easy and inexpensive. Smart, or "intelligent," terminals are those that can perform a variety of functions, including the storing and organizing of information for transmission on the network.

The intelligence of a terminal can be easily increased. For example, adding a chip to the telephone so that it remembers the last number dialed and providing a button on the keypad for calling up that number helps transform a dumb terminal into an intelligent one. Indeed, many cordless phones have this capability and are advertised as intelligent telephones.

Television receivers are, essentially, dumb terminals. While they permit us to obtain information and entertainment in often vast quantities (and, some might add, for that reason alone they ought to be thought of as dumb!), they are single purpose and respond solely to being turned on or off and switched from one channel to another. Add a device that allows you to inform a person or an intelligent terminal at the transmitting site that you wish a specific program or that you do not think very much of what you are receiving and you would have an intelligent television set.

As techniques such as frequency and time multiplexing are developed to send many signals along the transmission lines which enable us to use valuable bandwidth so efficiently (see Chapter 2), it is necessary that signals be properly "prepared" for transmission. The variety of uses of the network—voice, data, facsimile, video, and the special forms of each—makes this preparation difficult for the network to accomplish and still be flexible enough to handle many different types of signals. Hence, signal preparation has become the task of the terminal, although networks, too, have become more intelligent.

INTELLIGENCE ON THE NETWORK, TOO

No longer can you expect that a telecommunications network will interconnect the same terminals and transport the same mode of signals such as voice or data alone. Today's networks are designed to be *transparent*—that is, they accept information from a large number of terminals transmitting and receiving many different kinds of signals. The telephone network has always been a network transparent to all kinds of telephones and other equipment that could operate within its bandwidth limitations. In part, this was possible because all or most of the equipment used was licensed by the Bell Company, and if not manufactured by Western Electric, was at least designed to follow Western Electric specifications. Bell system network protocols were adopted by all independent telephone companies to ensure their interconnection with AT&T's long-distance services. Network protocols are designed for this very purpose—to allow networks to communicate between many different types of terminals and other networks.

Intelligence is also incorporated into networks and here again is an example of the tradeoffs we so often find in telecommunications system design. For public telecommunications systems such as the telephone network, intelligence is incorporated directly into the network, thereby allowing for low-cost terminals that can be available for wide public use. Without a telephone network made transparent by the high intelligence built into it, universal telephone service at affordable cost would not have been a feasible national communications policy. (In our chapter on the telephone, which follows, we shall explore this highly intelligent network.)

For some of the more specialized communications functions—functions more likely to be used by fewer people such as private lines that interconnect geographically dispersed offices of large firms or systems likely to be used for data and video—intelligence is often required and designed into terminals to ensure that network protocols are followed. Generally, users of these specialized networks value highly the services delivered, hence the higher cost of intelligent terminals.

The ability to distribute intelligence—memory, processing, and other computer functions in transmission systems—coupled with the ease with which dumb terminals can be made more intelligent has led to the transformation of the nation's telecommunications infrastructure. With the rising demand for the public telecommunications network (our telephone system) to exchange information among greater varieties of terminals, the network has had to become more flexible and more intelligent or face the loss of the highly valued information business. The public system's dilemma is how to become both more flexible and more intelligent without raising the cost of plain old telephone service. In the next chapter we will see how Ma Bell does so by improving her network intelligence. This, of course, has not stopped competitors from adding intelligence both to portions of the network and to the terminals used to deliver value-added services on networks that have become known as *value-added networks*. These new carriers have increased value to the existing network through the addition of intelligence on both the network and in the terminal.

THE MANY TERMINALS IN OUR LIVES

After the semiconductor chip and the communications satellite the terminal is the third fundamental tool of the communications revolution. It is the tool that allows us to create the communicating networks that are shaping modern telecommunications.

The odometer in your car is a terminal as are the new light displays that tell you your fuel efficiency at every mile. Terminals surround you in your kitchen as gauges on pressure cookers and as temperature displays on ovens. Terminals line the walls of your living room in the form of hi-fi turntables, cassette players and speakers, and radio and television receivers that put you on the receiving end of information as music and images. The revolution in communications that has brought us to this information era is most evident because of the terminals we use to communicate. Not only has there been and continues to be a veritable explosion of these devices, but they no longer determine what, where, when, and how we communicate. Rather, we choose to communicate what, where, when, and how we wish.

Today, the word "terminal" covers so many devices that they can hardly be recognized as members of the same family. Their "IQs" differ widely from the dumb to the superintelligent. They can look like typewriters, television sets, or work stations that include word processors, electronic mail terminals, printers, and other instruments of the office automation revolution. They can receive input through typing, sketching, or touching. There are terminals that can read your handwriting and some that can be taught to understand your voice. A few can even talk back.

Because the terminal is required for people to communicate electronically, designing terminals that are user "friendly" has been a major design challenge. In this case, the term "friendly" simply means that learning to communicate with the equipment is quick and actual communication is easy. On the whole, many users believe that the designers have not done well. Indeed, it has been remarked that the telephone was designed by engineers for engineers. It is at this point that the human factor becomes a concern in terminal design and configuration.

MAN AND MACHINE

A tension has always been present between man and machine. In the past, human factor analysts—known as ergonomic experts today—have generally concentrated on the factory floor, where the major investments in machines were made. When the typewriter, the adding machine, and telephone switchboards were introduced into offices, representing the first wave of today's much larger investments for office equipment, attention focused on the needs of the chair-bound information worker. While in the 1920s, human factor experts tried to adapt people to machines, today's ergonomists try to adapt the machines to people, to make information tools friendly and healthy. In some European countries, unions have already begun to campaign for special work rules for those who must stare into a computer screen for hours, and questions have arisen about the damage that might be caused by radiation coming from the video display terminal (VDT). Even in the absence of clear evidence, terminal designers are concerned about the potential for harm from the very low-level radiation emanating from VDTs over long periods of time. Firms wishing to expand their use of word processors and communicating computer terminals must be concerned with these issues and always on the alert for new findings that will reduce the tension between man and machine.

Other barriers to the use of terminals often arise. Why is it that so many professionals say they will use on-line databases but few actually do? Is it their fear of "computer illiteracy"? How important is their lack of keyboard skills to this fear? Is it that executives will not use keyboard terminals because they cannot type?

There are probably a great many more subtle reasons why terminal-accessed information is greatly underutilized—poorly constructed electronic information libraries, complex and unfriendly protocols that make "getting into" the libraries difficult, and the high cost of communicating with these libraries. But the terminal is critical; without it, interaction cannot begin.

The key to today's terminal systems, whether in the office or at home, is user interactivity. After all, the human user can, conceivably, add the best intelligence to communications. The success of a new terminal is increasingly affected by the quality of the user interface. As a purchaser or soon-to-be purchaser of telecommunications-computer equipment for your home or office, you desire a "user-friendly" system. The advertisements about various systems go to great lengths to convince you that a particular work station or personal computer will be just that. For some time, there have been rumors that the ultimate friendliness—the ability to by-pass the keyboard by talking to your computer—is coming soon. You will be able to teach your computer to recognize even the most peculiar idiosyncracies in your voice. How friendly can a computer eventually become?

You demand friendliness, for it will cut training time and reduce errors in the use of the network, and you and your staff will be comfortable with the machines. What more can you ask?

THE FRIENDLY TERMINALS OF TOMORROW

User friendliness can be achieved through the design of both hardware and software. The most significant development in today's terminals is that they are programmable, just as computers are. We can expect that more and more telephones, word processors, microwave ovens, televisions, and radios will become programmable in order to make the user's task even easier.

Not only do you want to adjust the microwave oven to the proper temperature for a particular dish, you want to start cooking at a prescribed time even if you are not at home. You can program the oven to temperature, cooking time, and start time, all by means of an intelligent device either in the oven or by the program you have stored in your intelligent telephone and which you call from your office.

In the office, your word processor is a remote terminal for accessing a computer in another department to search for information and perform the analysis you need for the mail you are sending across the country. Is your word processor simply a terminal or has it become a computer all on its own?

Is your new telephone, by which you can dial up the "Electronic Yellow Pages" for the latest product information at Sears, purchase it and tell your bank to pay for it, a terminal or a computer?

It is becoming more and more difficult to tell a terminal from a computer. The main reason for the confusion is that we can so easily increase the intelligence of even simple devices through the introduction of semiconductor chips that have the ability to remember what we tell them and perform according to our instructions. Rather than build in only one set of instructions, we can make the terminal capable of accepting many different kinds of instructions—that is, we can make it programmable.

Modern telecommunications networks transmit data in the form of conversations between computers, pictures, print, music, or whatever is necessary for information to be communicated. Such communication so often requires that the data or information be transformed during its transmission. This raises the all-important issue of what is communications versus what is computing. This is a very real concern because the communications industry has been a regulated industry while the computing industry has not. So who regulates what, if anything?

The multitalented chip started the debate in the 1960s and the equally talented terminals and switches made possible by that chip did much to resolve the argument in favor of an essentially unregulated computer-communications industry. This debate does not solely focus on the U.S. Chips and talented terminals have emerged in all of the developed nations of the world. Japan, Great Britain, and France, are three of the advanced countries with nationalized telecommunications monopolies that are now being forced to reconsider their monopoly operations. As the developing countries enter the information era, joining satellite networks, purchasing or manufacturing computer equipment and systems, and restructuring their telephone networks, chips and talented terminals are reshaping traditional industry structures and government institutions.

ADDITIONAL READINGS

Terminals are rarely, if ever, formally considered one of the tools of the communications revolution. Discussions of networks and their architectures are usually found in texts having to do with data communications and as cursory reviews in chapters dealing with telephone or broadcast technology. We shall have more to say about terminals, networks, and their architectures in subsequent chapters on the equipment that use them. In the meantime, here are several sources that you might wish to examine for more information on terminals, networks, and architectures.

Bernstein, George B., and Arnold S. Kasha, *Intelligent Terminals, Functions, Systems, and Applications* (Wellesley, MA: QED Information Services, Inc., 1979). One of the few complete discussions of terminals of all kinds, intelligent and not-so-intelligent.

Dordick, Herbert S., Helen G. Bradley, and Burt Nanus, *The Emerging Network Marketplace*, 2nd ed. (Norwood, NJ: Ablex Publishing Corporation, 1985). See Chapters 3 and 4 for discussions of terminals for network information services applications.

Martin, James, *Future Developments in Telecommunications*, 2nd ed. (Englewood Cliffs, NJ: Prentice-Hall, 1977). See Chapter 19.

PART THREE

THE TELEPHONE: NO LONGER PLAIN OR OLD

These are the tools of the communications revolution: the multitalented semiconductor chip, the communications satellite, and networks and terminals. These tools will launch the second industrial revolution and are the embodiments of the postindustrial society. We have already seen how they have altered the nature of work in an information society, and we have evidence of how they might change national economic, political, and cultural institutions and their potential for impact on a global scale.

In the chapters that follow we examine how these instruments are used in modern telecommunications systems. For it is in how they are applied to the tasks that need to be performed to enhance human communications that the power of these tools will be most evident. We begin with the electronic communicating instrument that is seemingly everywhere, the telephone.

CHAPTER 6

TELEPHONE TECHNOLOGY: TAKING THE MYSTERY OUT OF THE COMMONPLACE

HOW MUCH OF A DIFFERENCE HAS THE TELEPHONE MADE IN OUR LIVES?

Would it surprise you to know that we cannot really answer this question? Would you believe that this remarkable instrument has been all but overlooked by social scientists and historians? They have traced the impacts of the steam engine, the railroad, the printing press, and the automobile, but not the telephone. Volumes have been written about the cinema and its effect on families, cities, and nations, but very little has been said about the telephone and its impact on our lives.

We have largely taken the telephone for granted. Yet this wonderful toy—Bell's electrical toy as it was called when it first appeared more than 100 years ago—grew remarkably quickly in popularity as the nation emerged from gangling adolescence to maturity and leadership as the wealthiest nation in the world. The telephone must have had something to do with our nation's development. Surely it has not been a bystander. But exactly what was that something?

The desire to communicate at a distance is as old as legend. Abraham, Moses, and, in fact all of the Hebrew prophets, held frequent conversations with a God who was present only in spirit, much to the consternation of their friends, families, and neighbors. The Greeks, somewhat less imaginative, required their gods to appear from time to time and often in odd forms. Prior to the invention of the telephone, literature was cluttered with lovers who experienced each other's presence while separated by hundreds, even thousands of miles. Heroines held their breath for weeks awaiting their hero's arrival, spurred on by Jungian events of synchronicity, perhaps the most sophisticated tool for communicating at a distance.

So it should come as no surprise that at least five telephone inventors were racing each other to the patent office; Bell won by thirty minutes over Gray. But what happened after Bell spilled the acid *was* a surprise. Intuition tells us that the telephone and the telegraph would satisfy your need to communicate and, while not replacing a lover's kiss, they would reduce less romantic contact and travel. The evidence, however, points to the fact that all communications designed for interaction and transaction, of which the telephone is one, seem to affect society in counterintuitive ways. Consider, for example, that the number of trips made increases in direct proportion to the number of telephone calls placed.

Wherever we look, the telephone seems to have affected us in diametrically opposite directions. The telephone was thought to save physicians from making house calls, but physicians initially found that it increased them; patients could summon the doctor to their door rather than travel to his or her office. The phone invades our privacy with its strident, insistent ring but it also protects our privacy by allowing us to transact business way from home. It allows the dispersal of centers of authority, but it also provides for tight, continuous supervision of field offices at a distance. The telephone makes information available, but it reduces, and often eliminates, written records that document facts.

Before the telephone was invented three-fourths of the U.S. population lived in rural areas. One hundred years later, in 1976, only one quarter lived outside of metropolitan areas. Improved, mediated human communications were supposed to allow for the dispersal of population, not for population concentration. Yet John J. Carty, the Bell System's chief engineer in the early years of this century, argued that skyscrapers would not have been possible without the telephone. He said

> ...take the Singer Building, the Flatiron, the Broadway Exchange, the Trinity, or any of the giant office buildings. How many messages do you suppose go in and out of those buildings every day? Suppose there was no telephone and every message had to be carried by messenger. How much room do you think the necessary elevators would leave for offices? Such structures would be an economic impossibility.[1]

Today there are some 400 million telephones in more than 200 countries operating on what is essentially a single global network. Annually, this network carries about 400,000 million conversations, and they are increasing at the rate of 20 percent per year. By the end of this century we expect that there will be 1500 million telephones and more than a trillion calls each year.

Today almost 98 percent of U.S. households have telephones. Less than one person in ten would give up the phone if financially strapped. As the cost of long-distance services has decreased, an enormous increase has taken place in long-distance calling, indeed, in worldwide calling. Families on welfare are allowed to use their allotments for telephone service and senior citizens get special rates. Telephones truly provide "lifetime" services. When a disastrous fire on February 27, 1975 destroyed a major switching center in Manhattan leaving a 300-block area without phone service for 23 days, disconnecting almost 145,000 telephones, and disrupting service for nearly 100,000 customers, the cries of anguish were heard throughout much of the nation. There was no satisfactory alternative to the telephone; there was no substitute for the immediate interaction the telephone could provide. And serious business losses were the result.

The uses to which the telephone can be put are more varied and more numerous than most of us would have believed. We are all aware of how important the telephone is for making appointments, obtaining information on-the-spot, and managing personal and business affairs. It is probably the executive's most important management tool, extending his or her reach far beyond the geographical limits of the office and second only to face-to-face interaction for effective direction and motivation. We also are aware of how important the telephone is for the student and researcher, assisting them in tracking missed assignments, and following up on recommended readings and information sources.

These are the telephone's instrumental uses, or uses for which an economic payoff can be rationalized as in time and transportation costs saved, or tasks performed which otherwise would not have been performed. The telephone also has intrinsic uses, or uses for which there may be no economic value but which just make us feel better. Gossiping on the telephone is sometimes necessary for relieving personal stress. Calling up a loved one to determine that all is well or to just keep in touch, coordinating a community affair on behalf of a volunteer organization, and polling voters on issues are all intrinsic uses of the telephone that are often overlooked when pricing telephone services and determining if nations should receive loans for telephone systems development.

We are only now becoming aware that telephone services are differentiable products; not every use of the instrument is of the same value to the user.

The telephone belongs to that unusual class of instruments that affect society in ways determined by society. The telephone is a facilitating device; the consequences of its use depend in great part on purposive human calculation as do economic behavior and games. If we want to substitute communications for transportation, we can do so. If we want to use the telephone to work in skyscrapers, we can do so. We can choose to do with the telephone whatever it is we want to do with the telephone.

For these reasons and more, each of us should know all there is to know about Bell's wonderful toy, about the nineteenth-century luxury item that has become a necessity in the twentieth, and about how it will likely become the global lifeline of the information age.

TELEPHONE TECHNOLOGY

Three elements make up the telephone system: the subscriber loop or, as it is often called, the local loop; the switching system where so many of the new "chip" developments are reshaping the telephone; and the long-distance or line-haul system.

The Subscriber Loop

In this section we describe the functioning of the local loop. We discuss handsets, local lines, local loop signals, and pulse and tone dialing, all of which you use every day of your life and which in the coming years will be your entree into the information age (see Figure 6-1).

There are four local loop functions:

Voice transmission—As we found in Chapter 2, signals are transmitted by varying the power or amplitude of a direct current in the connecting lines or the transmission line in accordance with the sound pressure that is produced at the handset microphone. You will recall that this is amplitude modulation and, further, that this is an analog mode of operation.

Dialing—The direct current is interrupted by the dialing mechanism, whether this mechanism is for tone dialing or rotary dialing. This is a form of pulse modulation.

Ringing—Alternating current is applied at the switchend of the line thereby ringing that awful bell. The bell stops ringing when the handset at the ringing phone is taken off the hook, or goes *off-hook*.

Off-hook—When the handset at the receiving phone is picked up, a switch closes causing the current in the line to activate the microphone in the handset. At the receiving phone, the amplitude modulated carrier with its signal is heard. If you are the calling party, however, you will hear the familiar dial tone when the handset is off-hook.

The four local loop functions are illustrated in the wave-forms of Figure 6-2.

If you use pushbutton rather than rotary dialing, the dialing signals are bursts of audible tones. These tones consist of two frequencies for each button and are assigned as shown in Figure 6-3. When pushbutton dialing is used, it is necessary to add special tone decoders at the local, or first-level, exchange in order to identify which buttons have been pushed, and thus which number has been dialed. Two tones are used for each number so that each is uniquely identified by the intersection of a row and a column tone, or frequency.

Through the local loop comes access to the worldwide telephone network. The local loop is the responsibility of the local telephone carrier to whom you send your monthly check and where you probably lodge most of your com-

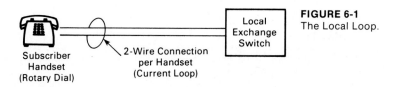

FIGURE 6-1
The Local Loop.

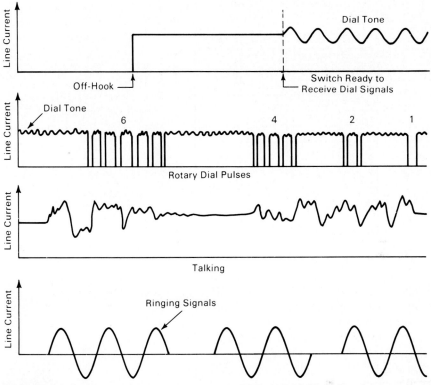

FIGURE 6-2
Managing Your Calls: Local Loop Signals.

plaints. Without this local loop, access to the information world would not be possible.

Until relatively recently it was believed that the local loop was a staid and rather stodgy technological wasteland; no one was very much interested in competing with Ma Bell and her sisters in that market. After all, the cost of wiring each and every household in each and every village, town, and city is enormous and the return on such an investment requires many years and much patience. And, as we noted in Chapter 2, there isn't much money to be made in transmission. Or is there?

The recent court decisions that led to the restructuring of the nation's telephone monopoly have let loose a surprising number of competitors for the local loop market. Firms offering several technologies including cable television, cellular radio (or mobile telephones), and a number of broadcast modes are finding it profitable to provide local loop services for special applications. For some years these were referred to as the "by-pass" technologies, a term that seemed to imply that any effort to compete with local telephone company services is, somehow, un-American. Perhaps a more appropriate way to look at

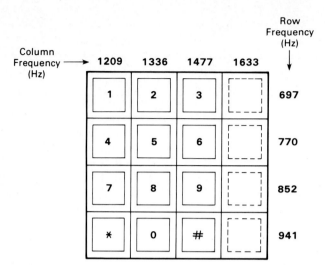

FIGURE 6-3
Frequencies on the Pushbutton Dial Keyboard.

these new local loop modes would be as technologies for segmenting the local loop market. We shall discuss these alternatives for local loop services in the next two chapters and again in Chapter 11.

Switching

In Chapter 2 we described the high costs of point-to-point networks; we observed that a switched network uses shorter copper wires or cables, thereby conserving expensive bandwidth. But we noted, too, that instead of many small, perhaps low-cost switches, the switched network requires a large, often complex, and certainly expensive central switch.

Since we are going to spend considerable time on the switching function, let's begin with basic functions, which do not change even as the ways they are performed radically do.

There are six basic switching functions:

To detect the telephone off-hook, to tell the system when the telephone is lifted off the hook either to make a call or to receive a call;

To connect the pulse receiver at the central office so as to be able to determine what number is being dialed.

To apply the dial tone to the line in order to signal the caller that he or she can now place the call;

To make the connection to the line specified by the dialing signals. How this is done depends on the type of call made which, in turn, determines the type of switch to be used. If the call is a local call—that is, if the parties are in the same exchange—the connection is applied directly to the called subscriber's line. If the call is to someone in another exchange, it may have to pass through intermediate switching centers. But in the end, the local exchange will connect the call to the called parties;

To ring the called party, if the line is not being used. If the line is busy, a busy signal is generated;

To detect that the call has been completed. When the phone handset is placed back on-hook, the switch breaks the connection. The smart switch also arranges to record information about the call for billing purposes.

Telephone progress can be measured in terms of switching progress for as new developments in switching technology are made available to the system, new services are offered. Indeed, the history of switching, from the step-by-step switch that replaced the manual operators who were the human controllers of switching during the early part of this century to the digital switches which were first installed in the mid-1960s, is the story of the development of microelectronics and computers. The birth of worldwide telephone networks also can be tracked by switch developments. And some might argue that the satellite is nothing more than a switch in space.

There are two types of switching—sequential and common control switching. In a *sequential switching system* connections are made progressively through switch paths as each dialed digit is received. For example, if the first dialed number is a 4, the switch moves four steps before selecting the connection corresponding to "4." If the next number is 5, it goes through the same procedure and makes the connection for "5." The next number dialed is, say, 4, and the switch moves another four steps and so on until the entire seven-digit number is dialed, signaled, and connected. In dialing this way, valuable switch and line capacity are used to complete a connection to a subscriber telephone that might, in the end, be busy. Switch capacity is expensive as are lines or trunks, so a means must be found to conserve switching resources. This is accomplished by the second mode of switching, called common control switching.

In *common control switching,* a portion of the switch, called the control segment, records the called party's number and by means of a separate signaling path called the common control path, checks to see if the called number is busy before making any connections. If the called party is not busy, the switch looks for the best switching path available at the time and makes the connection through that path. The common control path is, simply, a wire or cable along which control signals can flow. These signals can carry a great deal of information, information that can instruct other parts of the switch and, indeed, the telephone system, to perform all sorts of tricks. Thus, for example, modern telephone systems can set up call-forwarding, call-waiting, and call-storing services, teleconferences, and many other new services without tying up expensive lines used for talking. We shall discuss these new services in Chapter 8 on very intelligent telephones.

Switching functions have been performed in many ways over the years. Prior to the 1920s, women seated at long rows of plug-and-jack switchboards operated the telephone system, responding to requests for connections by plugging cords and jacks into switchboards. Today, switchboard operators are no more (or at least they are relatively scarce). Those who have recently become operators hold the more interesting jobs of responding to requests for such special services as reverse charge calls, police emergencies, crank calls and others that years ago were reserved for supervisors.

In telephone economics when a new switch becomes available the old ones are not simply thrown out; rather they are replaced when accountants say they have been fully depreciated. Consequently, there are still a few plug-and-jack switchboards in use today in the United States. In the main, however, you now find a mix of switches, from the step-by-step to the computerized common control, taking over. Let's discuss each of these systems.

Step-by-Step Switching Consider the fate of women had it not been for the step-by-step switch. Telephone managers estimated that a single operator with eighteen pairs of jacks could answer any of about 120 subscribers and connect them to as many as 10,000 subscribers. If telephone companies still had one operator for every 120 telephones, with more than 100 million phones in the nation, at least 800,000 operators, or 2,400,000 on three shifts, would be required. Of course with the equal opportunity efforts of the past several decades, at least half of these operators would now be men! It's almost as if someone was aware of this possibility and so created the step switch to take the place of almost 2.5 million switchboard operators.

Step-by-step switching systems were installed starting in the 1920s and are still operating in many parts of the nation. These systems perform sequential switching exactly as described above. Dial pulses directly control the relay that steps the switches along, selecting one of 100 connections per switch step. From each step single wires connect to local telephones. These switches were (and continue to be) in use primarily in local exchanges, the exchanges into which your telephone instrument connects directly. Local exchanges are the first exchanges your telephone call encounters; they are the Class 5 exchanges or switches we encountered in Chapter 5 and in Figure 5-1.

Crossbar Switching It wasn't too long before the slow step switches could not perform all of the tasks the rapidly growing telephone system required. With its two wires, the simple step switch could connect you to another telephone if that telephone was in the same switching exchange as yours. But if you wanted to call out of your exchange or make a long-distance call which required you to go through several exchanges, the step switches couldn't do the job. As the demand for long-distance service grew, the circuits were often crowded while callers waited their turn to be connected through several switchboards once a switch path became available.

The crossbar switch was introduced in the 1940s to meet telephone user needs. The crossbar switch is what engineers call a matrix-type relay with four wires connected to it as shown in Figure 6-4(a). Note that with the four wires arranged in the matrix form there are four possible connecting points rather

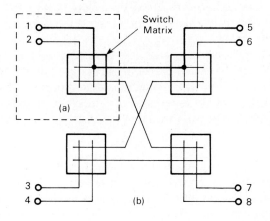
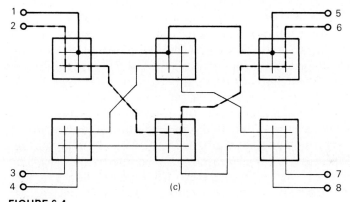

FIGURE 6-4
The Crossbar Matrix Switch.

than the one that is available when only two wires intersect. Furthermore, their arrangement allows for more than one path through the switch.

The term crossbar comes from the horizontal and vertical bars used to select the contacts, or crosspoints, connecting the signal path through which the parties talk once an open line is found. That is to say, you don't actually talk through the crosspoints but over the line they set up for you.

When the telephone is returned on-hook, the crossbar circuit is opened and the crossbars are released to look for other open lines to interconnect. While the crossbars search for open and available connections for your call, they are, in effect, storing the number you called until a connection is found. This is the key to common control signaling, a technological trick that has done more than any other to modernize your telephone.

Add three more matrix relays as shown in Figure 6-4(b) and an additional twelve crosspoints become available. Suppose we are at point 1 and are talking to someone at point 5. The person at point 2 is trying to call someone at point 6 but cannot be connected because there is only one path from (1,2) to (5,6). Any one of these parties calling each other would *block* the other party.

Add another switch stage, or pair, as shown in Figure 6-4(c). Now there are eight more crosspoints and, consequently, alternate paths for calls. By way of the dotted path, we can now get from 2 to 6 even if parties are talking on the line connecting 1 to 5. The addition of these switch stages results in a *nonblocking* condition, one of the many benefits of the crossbar switch. Nonblocking means that more calls can be put through and fewer nerves are frayed simply waiting for a dial tone.

Electronic Switching In 1965, not long after the invention of the transistor, electronic switches were introduced. The switches themselves were not actually electronic or even transistorized; they were, however, controlled by computers or equipment that performed computer-like functions. Indeed, the very early so-called electronic switches were computer-controlled crossbar switching systems. Only later were the electromechanical switches replaced by solid state devices. (Refer back to our semiconductor switches of Chapter 3.)

Remember the ability of even the simple crossbar switch to store the numbers you were dialing until an open connection was found? With the computer the control portion of the switch could now accept and store a great deal of information dialed into it and deliver this information to the switch when the switch is ready to make a connection. These switches were able to work faster and much more efficiently in finding connections for callers. In this way the telephone industry was able to greatly increase the speed of calling and, consequently, the number of calls that could be handled without a major increase in the number of switches or operators. This is an excellent example of how the computer can significantly increase the productivity of a particular industry by doing more with less. In a nutshell, it illustrates one of the more important issues facing an information society—the impact of the computer and the semiconductor chip on employment.

Common Control Switching The crossbar switch with its matrix format is the basis for today's modern common control exchange. The switches need not be electronic; as noted previously, they can be and often are electromechanical. Many times, the switch is simply several banks or stages of crossbar switches of our familiar matrix design controlled by microprocessors performing computer functions. Figure 6-5 offers a look inside a typical common control exchange.

Together, the concentrator, common control, link, register, sender, and scanner make up what can be called the computer in the exchange. When a call is made—that is, when an instrument in the local loop goes off-hook (Figure 6-6)—the scanner detects the signal and sends it to the common control. The common control searches for a free register—a place to store numbers—and when it finds one, allows the dial tone to be generated, thus recording the called number in the register. This is shown by the dotted line in

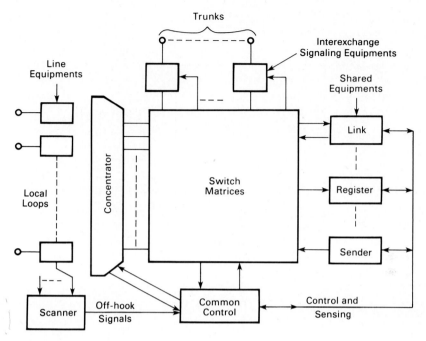

FIGURE 6-5
A Typical Common Control Exchange.

Figure 6-6. After checking that the called number is not busy, the common control searches for open switch paths. As soon as one is found, the register releases the stored request into the link and the conversation can begin. The register, so important in initiating the search procedure, is released for future searches as shown in Figure 6-7.

Common control allows for multiple actions to take place simultaneously; while one party is searching for an interconnection, another party may still be talking through the very same paths the waiting party will use when these lines become available. This makes for efficient use of very expensive switching equipment and also conserves copper cables since the calling parties do not have to wait on the line only to find that the called number is busy. Computer functions in telephone switching systems, functions that can remember and store information for later use, are the bases for today's *stored program control systems* or SPC switch.

Let us reflect for a moment on the evolution of the modern intelligent switch. An operator at a switchboard once provided the intelligent control needed to route telephone calls, remembering the numbers dialed, storing that information in the human brain until a line was open, inserting the plug and jack, and completing the call. In this respect, the operator provided control to make the system work as efficiently as humanly possible.

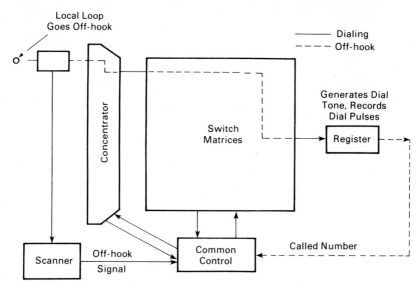

FIGURE 6-6
First Stages of Call Initiation.

Step-by-step switching did away with the intelligent human operator by combining the intelligent control function with the switching function. The result could have been, and usually was, a sharp increase in either the number of busy signals or in the amount of copper necessary for the added transmission lines and many more switches required to alleviate the problem.

The common control switching system separated the switching function from its control function through the use of special signaling paths and intelligent electronics, microprocessors and memories, storage devices, and other computer functions.

Stored Program Control and Digital Transmission You have heard much about the convergence of computing and communications. We now come to a prime example of this phenomenon. Most of us have always believed that IBM is the largest manufacturer of computers in the world and that an IBM, Univac, or Control Data model is the largest computer in operation. We are wrong on both counts. Western Electric, the now-independent manufacturing arm of AT&T, is the largest manufacturer of computers; its No. 5 ESS is probably the largest computer in operation today.

Computers that have to talk to other computers have been largely responsible for developments in digital switching and stored program control. For computers to receive and send information to and from each other and to and from remote terminals requires high speed and reliable and accurate switching. The transistor has made this possible.

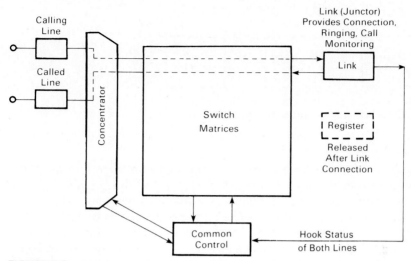

FIGURE 6-7
Final Stage of Call Initiation.

Electronics were first used in telephone switches to replace the electromechanical relays in the voice transmission circuits with transistors. This did not affect the voice transmission which was (and still is) an analog process. However, within the past several years there has been a surge of interest in digital transmission. New industrialized nations such as Thailand and, reportedly, China, have chosen to build digital transmission telephone networks for much of their new telephone infrastructures because they see the value of the system being able to provide voice transmission as well as high-speed digital transmission for computer communications. Networks that can provide voice, data, facsimile, or image, and even video transmission services on a single network are called *integrated services digital networks*. How digital systems work and why they are seen as the most efficient systems for communications will be discussed in the next chapter.

Line-Haul Transmission

We now come to the third major subsystem of the telephone network, the line haul, long-distance or trunk system. Line haul refers to the transmission paths required to carry subscriber-to-subscriber conversations between local exchanges to which all subscribers are connected by local loops as shown in Figure 6-8. If two subscribers are within the same exchange—that is, they are within the same local loop—long-distance circuits are not needed.

In the early days of the telephone, line-haul circuits were similar to local loop circuits; they used individual wire pairs for each circuit. Soon, however, the economics of telephone networking became apparent to both subscribers

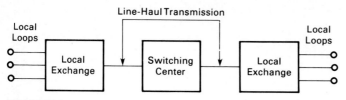

FIGURE 6-8
Line-Haul Transmission (Long-Distance Calling).

and providers. They both learned that the more subscribers on a network, the more valuable that network is to both the user and the provider. As the number of subscribers increased, especially in the years immediately following World War II, and as the number of telephone companies throughout the country sought to interconnect, demand for these long-distance wires soon resulted in intolerably long waiting times. A busy signal does not endear a customer to the telephone company. Consequently, some of the early research at telephone laboratories was devoted to increasing the capacity of long-distance circuits. Many of today's multiplexing techniques were initially applied to the line-haul networks and greatly increased the number of signal channels carried by the networks. So, too, new systems and hardware, such as microwave, satellites, and optical fiber transmission systems, were applied to long-distance transmission networks. A local loop circuit is used by a particular subscriber at any given time, but the trunks must be time-shared among many subscribers if the system is to interconnect a nation.

Line-haul circuits are interexchange networks. Because of their special uses, they differ from local loop circuits in two major respects:

1 There is no need to maintain compatibility with the handsets, and a variety of input and output signals may be used. The long-distance network is a "transparent" network; it cares little about the sending and receiving equipment attached to it.

2 The long distances that need to be covered by the circuits require that the signals be amplified at points along the lines. In order to perform this amplification and to alleviate some resultant quirks (echoes, for example) that accumulate when you send signals a long distance, long-distance circuits have four wires rather than only two as do local loop circuits.

What to Do about Signal Loss In Chapter 2 we saw what happens when a signal is transmitted—how noise and the characteristics of the material through which it is sent can cause distortion and reduction of the original power or amplitude of the signal. Over long distances—transcontinental, transoceanic, and now through space—these losses cannot only make the original message unintelligible but cause it to disappear altogether. As we see in Figure 6-9(a), the signal can fall below the level of noise in the system and be entirely lost.

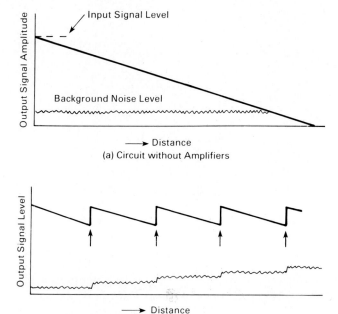

FIGURE 6-9
Signal Loss on Long-Distance Circuits.

To ensure that the signal will be greater than the noise, amplifiers must be inserted in the transmission line as shown in Figure 6-9(b). Often times, several are required. Note that the noise is amplified as well as the signal, so there must be a rather decent signal-to-noise ratio in the transmission line to begin with to receive the signal sent. Of course this can increase the cost of long-distance transmission considerably. However, the picture is different when the signals transmitted are digital signals as we shall see.

What to Do about Echoes When amplifiers are used for long-distance transmission lines, it is necessary that the signals travel in only one direction at a time, otherwise the amplified signal would crash into the original signal and you would get absolutely nowhere. Transmission lines with amplifiers must have two circuits or two pairs of wires—one pair in each direction—for two-way communications. Long-distance transmission lines are, thus, four-wire lines as compared to local loop lines which are two-wire systems because no amplification is required. To match the two-wire local loop circuit to the four-wire long-distance circuits, a device known as a *hybrid* is required. Figure 6-10 illustrates the hybrid's operation.

The hybrid separates the sending signals from the receiving signals. Sometimes it doesn't work as well as you would like and there is leakage between

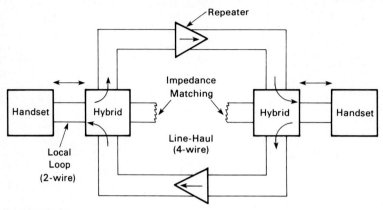

FIGURE 6-10
Local (2-wire) to Line-Haul (4-wire) Interconnections.

the sending and receiving portions of the hybrid circuit. Engineers call this coupling; you hear it as echoes. These echoes can be annoying and because of them, some heavy telephone users, especially business users, have refused to pay their telephone bills until they are removed. Echo removal is accomplished by the addition of *echo suppressors,* voice-actuated switches at each end of a four-wire circuit (see Figure 6-11). They operate on the principle that when one party is talking the other party is listening (or should be listening!) and does not need to talk until it is his or her turn to talk and the other party's turn to listen. How an echo loop is broken is illustrated in Figure 6-11. A problem with the echo suppressors, which you may have often noticed, is that they tend to clip the beginning portions of your speech and you may find that uncomfort-

FIGURE 6-11
Echo Suppression.

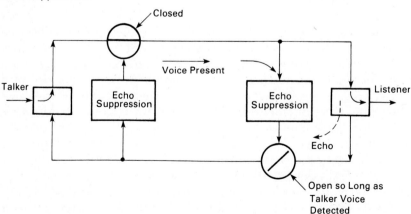

able. *Echo cancelers* are now being used which can be tuned to eliminate only the reflected signals rather than another party's speech. These cancelers store the transmitted speech for the short time it takes for the speech segment to come back to the talker, then play back what has been stored in opposition to the incoming echo, thereby canceling it out. Lo, the wonders of the intelligent microprocessor!

In recent years telephone line-haul systems as well as telephone switching have been the recipients of new technologies. In large part these, too, have centered on the semiconductor chip and computer developments, but there have been others. To appreciate these new developments it is necessary to understand the economics of transmission systems.

INTRODUCING TRANSMISSION ECONOMICS

In Chapter 2, we discussed the principles of multiplexing. There is frequency multiplexing by which a single transmission line is divided so that many voice signals (or data) can be sent at the same time. If you recall, this is done by imposing the signal on several different carriers, modulating the carriers, and separating the signals by frequency. In this way, it is impossible to send many signals on a single channel. How many different signals can be sent depends on the bandwidths of the signal and the channel.

In Chapter 2 we also showed how multiplexing can save money by using less bandwidth in low population areas and more bandwidth (or channels) in more populated areas (see Figure 2-15). The need to efficiently use bandwidth is especially critical in line-haul systems. Let's examine the economics of transmission in somewhat more detail.

Reviewing, there are two families of multiplexing methods: frequency division multiplexing and time division multiplexing. The telephone engineer also has the option of not using any multiplexing techniques but, instead, running one circuit for each communication channel. Which option to use depends upon the relative costs of transmission. In addition to the cost of the channel itself, the cost of the terminal equipment—handsets and the devices necessary to do the multiplexing—must be considered. These devices are required at both ends of the channel and, if their costs are less than the cost of the equivalent number of single-circuit transmission channels, then it is possible to multiplex and save money.

Since line costs are proportionate to the distance covered and terminal costs are fixed, multiplexing becomes more economic than individual circuits only beyond a certain distance. For a particular multiplex system, the minimum distance at which there is an economic advantage is called the *prove-in point*.

An economic analysis of multiplexing versus individual circuits is illustrated in Figure 6-12. Two multiplexing methods are compared with the individual circuit option. Let us suppose that multiplex system A uses frequency multiplexing and multiplex system B time division multiplexing. System A's cost per circuit mile becomes lower than the individual cost per circuit mile at a

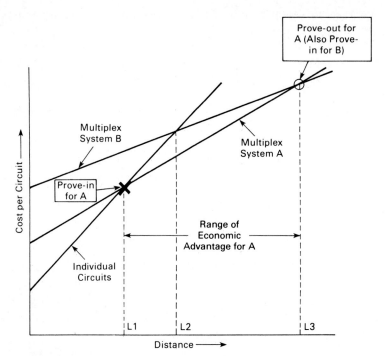

FIGURE 6-12
Choosing the Appropriate Circuits (Prove-In/Prove-Out).

distance of L1 miles. System B's cost per circuit mile does not *prove in* until a distance of L2 miles. System A has an economic advantage over both system B and the individual circuits up to a distance of L3 miles. At that point, it *proves out* in favor of system B.

This conclusion makes sense, for we can expect that time division multiplexing costs are likely to be higher than frequency multiplexing costs; for time division multiplexing, it is necessary to add the cost of coding and decoding and the costs of converting analog signals to digital signals for transmission and back again to analog signals so that the messages can be deciphered. What do you think will happen when the local loops go digital?

Dramatic reductions in circuit-mile costs have been made during the past fifty years, sometimes by as much as a ratio of 100 to 1. These reductions have been achieved by building larger and larger capacity systems. But lower costs can only be realized if there is a demand for many circuits on a particular route. In planning a voice or data communications network that includes long-distance as well as local loop transmission, the choice of the best technology for each part of the network is essentially based on traffic estimates for the various portions of the network. Because more choices of technology have been available for long-distance transmission than for local loop transmission, more cost savings have been achieved on the long-distance side. Long-distance costs have been steadily falling while, at least up to now, local loop costs have remained essentially constant. While costs have had a great deal to do with the regulatory practices of state public utility commissions who sought to ensure universal telephone services in their states for humane as well as political reasons, line-haul transmission technologies have also influenced costs. For example, the choices now available to system planners for line-haul systems are:

- Cable wire pairs—the old-fashioned local loop working over long distances,
- Coaxial cables—which were developed more than forty years ago and are used today not only for long-distance telephones but also for cable television,
- Microwave radio relay systems which have, for several years, been replacing the coaxial cable for cross-country transmission, but which are now being supported by satellites, and optical fiber transmission systems.

It is a truism of almost all technology, and especially so of the telecommunications technologies, that new developments do not entirely displace the old. We can expect that for many years to come all of the available techniques for providing long-distance communications will be economic in specific situations. Thus, for example, microwave guide systems with very large channel capacities in the neighborhood of 230,000 to 460,000 voice circuits are being developed in parallel with ever higher frequency satellites and optical fiber transmission systems. A sort of race is going on among these technologies and you can expect that all three will find their appropriate roles in telecommunications systems.

OPTICAL FIBERS IN YOUR TELEPHONE FUTURE

A special place for optical transmission systems is reserved in the telephone system. Optical transmission systems are being installed in several places in the United States and throughout the world. The reasons for the great interest in these systems are twofold: (1) optical transmission provides almost unlimited bandwidth and can deliver unlimited channel capacity; and (2) the glass fibers that carry these channels take up little space and are very inexpensive. After all, there is a great deal of silicon on the beach! It is expected that, in time, optical transmission systems will replace not only certain portions of today's long-distance transmission systems but also will find their way into the local loops and into our homes.

Using light as a means of communication goes far back in history. Remember the light signals that sent Paul Revere on his now-famous ride? Long before that, mirrors were used by armies to send reflected sun signals to each other from mountaintop to mountaintop. More than a century ago Alexander Graham Bell demonstrated the photophone, a device consisting of mirrors and selenium detectors that carried speech no more than a few city blocks. Nevertheless, he showed that intelligence could be transmitted on a light beam. All he needed was a better source of light and a transmission medium more reliable than the smoggy, foggy air of Boston.

The invention of the laser in the 1960s provided that improved light source, and in 1970, glass fibers capable of delivering light waves without interference or excessive loss were developed. The necessary components were now in place; all that remained to be found were the electronic components necessary for the transmission of signals. This required the redesign of amplifiers, terminals, multiplexing devices, coding and decoding devices, and other familiar pieces of communications equipment. Even when this was accomplished by the late 1970s, the techniques for working with glass fiber systems remained to be developed.

Optical fibers are thinner than a human hair. But because glass is a very strong material under tension, but brittle and easily cracked or nicked when bent too far, the fiber optical cable must be made up mostly of material to cushion the glass and to provide mechanical support.

Transmission by optical fiber is very similar to transmission by copper wire. The functions required to generate a signal and to detect and make intelligible the information transmitted are the same. The similarity of optical transmission to wire transmission is illustrated in Figure 6-13. What is unique is, simply, the cable itself. If the fiber is manufactured correctly, its refraction index (the ratio of the velocity of light in a vacuum to its velocity in the glass fiber) in concert with the surrounding material, will prevent light from escaping from the sides of the fiber and thereby keep losses down. This is exactly what is necessary to achieve efficient transmission.

Transmission by optical fiber is attractive for many reasons:

- Glass is more plentiful than copper;
- The small size of the cables helps to increase the capacity of ducts that are now crowded with conventional cables;

FIGURE 6-13
Fiber Optic Communications.

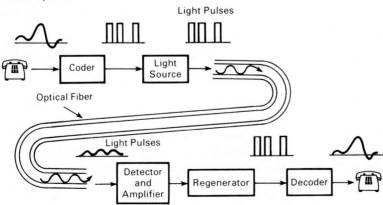

- All forms of electrical interference are eliminated;
- The bandwidth potential far exceeds that of electrical cables; and
- The low resistance or attenuation even at very high frequencies reduces the number of amplifiers required to deliver signals over very long distances.

THE FUTURE OF THE TELEPHONE

Theodore Vail, the organizational genius who made Bell a household word, had a "grand design" for a "universal service"—a single communications network that would interconnect a nation. Indeed, our nation is now interconnected by more than 1500 independent telephone companies acting as a single system and providing telephone service at "affordable rates"—that is to say. at rates that can be met by more than 98 percent of the households in the nation.

Vail's dream was given reality by the Communication Act of 1934 which made the Bell System a virtual national monopoly in order to provide universal telephone service at affordable costs. To merit special grace, the Bell System agreed, in a 1956 Consent Decree, not to engage in any services that could not be classified as transmission. The convergence of the communicating and computing technologies, which we see so clearly illustrated in modern telephone switching, blurs this distinction. As noted previously, information that is transmitted today can be manipulated, translated, stored, even modified before it is received in order to achieve efficiencies in communications.

We should recall, too, the several benefits of all-digital transmission systems: low cost per message because of efficient multiplexing; no need to provide hardware for analog-to-digital conversion and the reverse, savings in repeater hardware; and, finally, a single network for voice, data, video, and image transmission.

The consequences of almost ten years of debate came to a head with the Greene Decision of 1983 which deregulated AT&T, thereby allowing Ma Bell to enter into nontransmission businesses. In return for lifting the 1956 Consent Decree, AT&T agreed to dispose of twenty-two of its operating companies (the local Bell Operating Companies), and to set up a separate subsidiary for its nontransmission (read "computer") ventures. AT&T's manufacturing arm, the Western Electric Company, and its Long Lines Division will now be competing even more aggressively to provide computerized and other specialized terminals, with information hardware, systems, and software as well as with firms offering long-distance services by satellite and other modes.

Local loop service remains the responsibility of the local Bell Operating Companies, albeit in a new configuration of seven holding companies into which the individual operating companies have been folded. It is important to note that the operating companies remain regulated by state utility commissions which now have the sole responsibility for ensuring that telephone rates remain affordable, whether they deem this a social responsibility or a political necessity.

New technology to the rescue? This has, at least up to now, been the telecommunications industry experience. Already technologies have begun to emerge that offer the prospect of providing alternative local loop services, among them, cable television. If these so-called by-pass technologies can establish a viable market, they will offer competition in the local loop market and might keep the price of local services at affordable levels. However, this remains to be seen.

Some forecasters have suggested that there will be a single, all-digital network for all forms of communications including voice, data, video, and images. While technologically this is quite feasible, it is more likely that different services will continue to seek out different transmission means. The bank will look to the high-speed data services cable might offer, while the small business will continue to use the telephone to reach the bank. Airlines will use special radio equipment and frequencies quite different from those used by the radio listener in his or her automobile. Satellites will continue to be used for long-distance transmission while the twisted pair or the cable, and perhaps the optical fiber, will be used for voice and personal computer transmissions in the home. A variety of transmission modes will be used side by side because it is economical and useful to do so. Users will pay only for the bandwidth they need.

What about universal service at affordable cost? Who will decide what universal service means? Is universal service a dial tone in every home or access to information networks in every household? And how will services be made available to those who may not be able to afford costs that have risen, in some areas of the country, by as much as ten times?

These are questions of equity rather than efficiency and as such they are political issues rather than engineering issues. But the technological-political interface is inescapable; it is your task as an intelligent user of communications to act on this interface.

YOU AND THE TELEPHONE

This, then, is the telephone, that wonderful instrument on which we have come to depend and which promises to be even more important to us in the coming information age. We have described the basic telephone system as it has evolved over more than 100 years, an AT&T system that connects us to the world. This is POTS, plain old telephone services, and we have seen that it is not really very old and not at all very plain after all.

The telephone is, indeed, the most modern of communications systems. What more can one ask of a communications system than that it have the ability to instantly create communications suitable to a particular need. Your communicating skills may be your most valuable assets at home and in the office. The art of living is primarily the art of communicating with people and the telephone is a system that can be molded to your specific personality and need.

We have followed the evolution of modern switching and have hinted at the wide range of new services it makes possible. We have discussed the computerization of common control signaling in modern switching and we briefly mentioned its far-reaching impacts on the modernization of the telephone. It is, in fact, the single most important development in telephony and has opened the way for the modern PABX (Private Automatic Branch Exchange) and the era of modern voice services which we shall discuss in much detail in Chapter 8. To this discussion must be added a word about the digitization of the line-haul systems, a process which is proceeding at a rapid rate throughout the world and somewhat more slowly in the United States. As we saw in Chapter 4, the satellite is rapidly altering the shape of long-distance telephony and with it come many new communications services. And in the next chapter we shall more fully examine just how the telephone, although more than 100 years old, is now joining the information age. Just what this marvelous instrument can do for you is the subject of the two chapters that follow.

REFERENCES

1 This wonderful quotation is found in Ithiel de Sola Pool's *The Social Impact of the Telephone* which is highly recommended for additional reading.
2 Ibid.

ADDITIONAL READINGS

Literature on the telephone is, surprisingly, of quite recent origin. It seems that it took us more than 100 years to become fully aware of this powerful and wonderful instrument. There is relatively little on telephone technology other than the massive literature compiled by the engineers of the Bell Telephone Company. We have selected a range of additional readings from history to economics to technology.

Brock, Gerald W., *The Telecommunications Industry* (Cambridge, MA: Harvard University Press, 1981). An excellent economic study of the history of telecommunications in the United States.

Brooks, John, *Telephone: The First Hundred Years* (New York: Harper & Row, 1976).

Fike, John J., and George E. Friend, *Understanding Telephone Electronics* (Ft. Worth, TX: Radio Shack, 1983). This will provide those interested in more about telephone technology with some details.

Pierce, John R., *Signals: The Telephone and Beyond* (San Francisco: W. H. Freeman & Company, 1981).

de Sola Pool, Ithiel, ed., *The Social Impacts of the Telephone* (Cambridge, MA: MIT Press, 1977). A collection of papers prepared for the 100th anniversary of the telephone. This is the first serious examination of the social dynamics of the telephone.

de Sola Pool, Ithiel, *Forecasting the Telephone* (Norwood, NJ: Ablex Publishing Corporation, 1983). It's amazing how wrong "futurists" were about the telephone.

Talley, David, *Basic Carrier Telephony* (Rochelle Park, NJ: Haydon Book Co., 1977).

Talley, David, *Basic Telephone Switching Systems* (Rochelle Park, NJ: Haydon Book Co., 1979).

CHAPTER 7

INTELLIGENT TELEPHONES: MODERNIZING THE TRADITIONAL POTS

WHEN IS A TELEPHONE NOT A TELEPHONE?

The general manager of the department of public utilities in Los Angeles glared at the telephone on his desk. It had been installed over the weekend and this was his introduction to the device. The pictures he had seen in the catalog showed a friendly instrument; now, on his desk it seemed formidable. It wasn't a telephone, it was a—well, a terminal, a computer, a switchboard, a piece of heavy "artillery." It featured four rows of pushbuttons and lights, a speaker, a small TV-like screen, and—yes, the trusty handset. At least that looked familiar. But could this be a telephone?

He thought of the three black telephones he had on his desk when he joined the department forty-five years ago, one for each of the three telephone companies then serving Los Angeles. They seemed more friendly; all you had to do was juggle the three phones as the calls came in—no buttons to push, just pick up the phone and talk.

When one telephone could reach all of Los Angeles, life became simpler. There was only one telephone on his desk and a receptionist at a switchboard to juggle the calls. A few pushbuttons were later added to the phone and he had four lines coming into his office. He rather enjoyed telling a disgruntled citizen that he had a "call on the other line and would the good citizen hold?" He used the waiting time to think up a reply for that good citizen. Several years later, the switchboard was replaced by a neat console, with several buttons and lights instead of plugs and jacks, and the receptionist was able to handle calls a great deal faster. But the telephone on his desk was still a familiar, friendly instrument.

Now there is this "hunk of hardware" with its own instruction book! Imagine instructions for a 63-year-old executive who has used the telephone all his life! The young salesman who had briefed him earlier in the week talked "computereze." He spoke about a smart telephone that had the ability to remember the last number dialed and up to twenty frequently called numbers. All you had to do was press a button and the phone would dial the number for you. He said that with this intelligent telephone, you could set up a meeting with as many as six people no matter where they might be—in or out of the building—and at the push of a button they'd be on the line. When you left your office you could tell the telephone where you could be reached and it would follow you around, forwarding your calls to whatever number you programmed into it.

Did the general manager purchase a telephone or a computer?

This is the $100-billion question that has jarred the telecommunications industry in the United States and is causing shudders among telecommunications industries throughout the world. For this is the world communications equipment market of the future, and every major supplier of telecommunications *and* computer hardware is after a share.

In the previous chapter we saw how important computing is for communicating: microprocessors (the common control in the switches we discussed) remember numbers called and instruct switches when and how to forward conversations, thereby making more efficient use of valuable bandwidth (and copper). We will see in Chapter 10 on mobile radio how computers allow us to talk while on the run, keeping track of us as we move from neighborhood to neighborhood in the city. Computing is essential for modern telecommunications. It is not surprising, therefore, that the computer and telecommunications industries see that possibilities for growth lie in each other's territory. The consequence of that growth is an explosion of products flooding a marketplace that does not resemble the sedate old world of Ma Bell.

In this chapter we examine how computers are converting our friendly POTS into an intelligent, indispensable information instrument.

THE TELEPHONE AS AN INFORMATION TECHNOLOGY

Your communicating skills are your most valuable asset. The art of management is primarily the art of communicating so that you can achieve the results you want. Selling is communicating, and scientists who cannot tell anyone about their discoveries will certainly not win Nobel prizes.

The telephone is perhaps your most important instrument of information technology. Ask any teenager and you will find that the telephone is one of life's necessities and that an hour or two on the phone is more entertaining and informative than an hour or more of television. Political campaigning would be dead in its tracks without the telephone for those all-important popularity surveys and for reminding voters to visit the polls. Multiplant firms throughout the world would be paralyzed without the ability to distribute information easily and rapidly among their many offices via telephone. Despite the complexity, size, and massive capital requirements of the telephone system, it provides an instrument for human interaction that is convenient and easy to use. It is immediate, personal, and can be custom-tailored to specific needs. No wonder it has been called our most modern communications technology. Furthermore, the telephone is learning all sorts of new tricks. Let us examine what the general manager can do with his new telephone.

By following the instructions in his *Manual on How to Use Your Intelligent Telephone* the general manager can:

- Teleconference, with or without pictures, with or without printed material,
- Call and manage a conference with anyone in the world,
- Forward calls to wherever he might be, anywhere in the world, so that he is never out of touch,
- Automatically, at the push of a button, dial frequently called numbers,
- Store and forward messages and eliminate that perverse form of torture called "telephone tag,"
- Design a personal message-answering and message-delivery service,
- Be informed that a call is waiting without interruption of an ongoing conversation,
- And much more.

How did the plain old telephone become so smart?

THE ARCHITECTURE OF THE TELEPHONE SYSTEM

In Chapter 6 we discussed telephone technology and described the basic components required for a telephone system. Now we examine how these components are organized into the public switched telephone network, (PSTN), the core communication system of this nation and, increasingly, of every nation entering the information era. To do so we draw upon our discussions of multiplexing (in Chapter 2), semiconductors (in Chapter 3), and the architecture of telecommunications (in Chapter 5), a good way to review what we have discussed so far in our study of modern telecommunications.

We found that the star topology illustrated in Figure 5-1 is the basic architectural module or building block of the telephone system. The level 5 switch in that illustration is the central office through which your telephone is interconnected with other telephones for local services and with the line-haul system for long-distance services. At last count there were about 17,000 central offices or wire centers in the nation. In general, each central office of a telephone system in the United States has a maximum capacity of 10,000 lines and serves a geographic area called an exchange. (Telephone lore is full of stories—apocryphal but, nevertheless, interesting—about the reason for the 10,000 lines; it is said that they are about the length of an operator's span from one end of the switchboard to the other with both hands outstretched.) These exchanges are probably more familiar to your parents than to you; they knew them by names that often reflected some local neighborhood characteristic such as University, Plaza, Chestnut, Rittenhouse, or Walnut. These names were easy for human operators to handle but when automatic exchanges were introduced, two-letter abbreviations and numbers offered more choices for new exchanges, and hence the names became UN-1, PL-5, CH-6, RI-9, and WA-4. As the number of subscribers grew, additional exchanges were required and letters gave way to numbers—more combinations of three-digit numbers were possible and the common control computers managing these switches could do very well with numbers. How a central office serves an exchange area is shown in Figure 7-1.

In a metropolitan area there are many exchange areas requiring many central offices. These could be interconnected as shown in Figure 7-2. However, as noted in Chapter 2 in our discussion of switched and non-switched networks, point-to-point interconnection is very expensive in terms of both copper and bandwidth (see Figure 2-1). Just as we inserted a switch in our network in Figure 2-2 we do so again in the metropolitan area, with a tandem switch in a tandem office, shown in Figure 7-3. About 850 such tandem switches are located either in the tandem office or in newer offices called toll centers. Note that tandem offices complete calls between central offices but do not connect to subscribers.

The arrangement of tandem and central offices is a two-level hierarchy (see Chapter 5). As the number of subscribers continues to grow these tandem offices are interconnected just as central offices are linked. This interconnection could be done point-to-point as suggested in Figure 7-2 for the central offices,

FIGURE 7-1
Local Exchange Area.

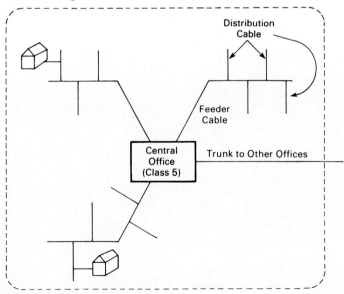

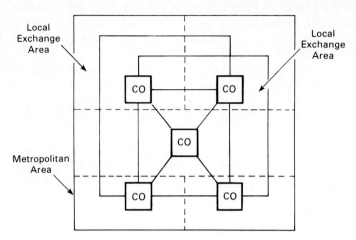

FIGURE 7-2
Metropolitan Area With Several Local Exchange Areas.

but that would require extensive cost outlays for copper and switching. So another switching hierarchy or level must be added. Actually three more levels are added so that the present PSTN becomes a five-level hierarchical structure (a considerably simplified version is shown in Figure 7-4). These additional levels or classes above the central office (class 5) and the tandem office (class 4) are known as the primary office (class 3 of which there are about 215 in the nation), the sectional office (class 2 with 65 offices nationwide), and the regional office (class 1 with 12 offices in North America). When the traffic between central offices or tandem offices is especially heavy, as is likely

FIGURE 7-3
Conserving Copper with a Tandem Switch.

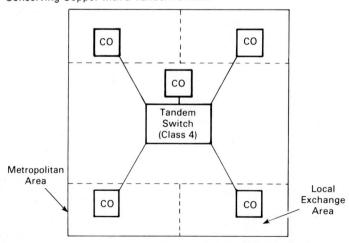

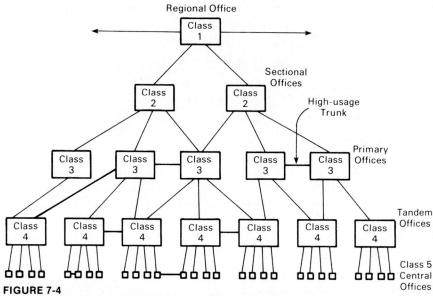

FIGURE 7-4
The 5-Level Public-Switched Telephone Network Hierarchy.

between central offices in highly congested business locations or residential areas, a trunk cable is used to interconnect them directly. This is referred to as a high usage trunk.

Saving copper is not the only reason for creating network hierarchies in the telephone system. It is necessary, too, to conserve transmission paths and still allow for the greatest number of simultaneous conversations without fraying the nerves of subscribers with too many "fast busy" signals. (The "fast busy" is what you hear when the switch rather than the party you are calling is busy.) We learned in Chapter 2 that one way of creating a *non-blocking* condition is to use the switch matrices shown in Figure 6-4. Another way is to provide a large number of alternate transmission paths by means of network switching hierarchies shown in Figure 7-4.

Think back for a moment on our discussion in Chapter 2 of multiplexing and, in particular, frequency division multiplexing (FDM). This is another way by which the number of messages on the twisted pairs of telephone wires can be increased. By combining the switching hierarchies of Figure 7-4 with the FDM hierarchies shown in Figure 7-5 we can create additional message paths, thereby reducing the probability that a caller will hear that frustrating "fast busy" signal. Indeed, at present there is only one chance in a hundred that a caller is likely to get "blocked." With the application of time division multiplexing (see Chapter 2), the blocking probability will be further reduced, even as the number of messages entering the system increases. In our next chapter, we shall discuss how time division multiplexing is leading the way to the telephone of the twenty-first century—the all-digital transmission system.

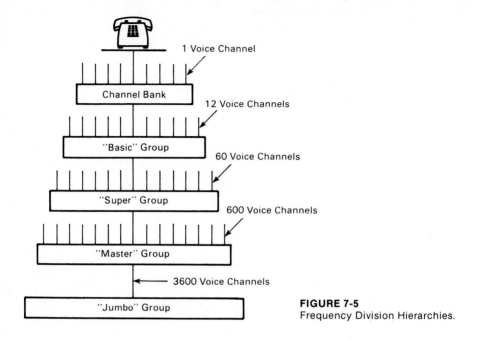

FIGURE 7-5
Frequency Division Hierarchies.

CONTROL SIGNALING AND THE IMPORTANCE OF THE THIRD WIRE

Control signaling is not new to the telephone system nor to us. We talked about control signals in our discussion of the *the subscriber loop* in Chapter 6. These are the in-channel signals heard by the subscriber:

• The dial tone that informs the caller that the system is ready to receive the number dialed,
• The signals that transmit the number dialed to the appropriate switching office,
• The busy signal that informs the caller to try again, and,
• The ringing that tells the recipient there is a call on the line.

These control signals take us from the subscriber to the local central office. A more comprehensive signaling system is required to integrate the rest of the network, one that can impart intelligence—memory and decision-making—to the telephone system.

In the early days of the telegraph and telephone, it was often necessary for operators and engineers to talk business with one another—for example, to inform connecting networks of scheduled downtimes, discuss and solve transmission problems and, as the communications on these networks became crowded, to manage transmissions by planning ahead. For just such purposes, a separate channel called an "order wire" was often used. Today this old order wire has reappeared as the *common channel interoffice signaling* (CCIS)

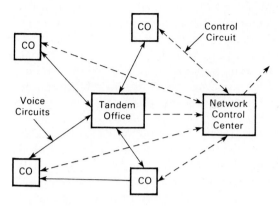

FIGURE 7-6
Common Control Interoffice Signaling.

systems in modern telephone plants throughout the world. A diagram of a CCIS system is shown in Figure 7-6; the dotted line is today's order wire.

The CCIS system does not have to be associated with any particular message or voice circuits; note that this version manages and controls the voice network by way of its own network control center and does not send its digital signals on the voice network. Another version of a CCIS does send signals on the voice network and just how it does so will be described in the next chapter.

So we see that with the arrival of stored program computer-controlled switches and the disappearance of human operators in offices, computers now talk to computers. The old order wire no longer transmits voice through the network; it transmits data. Signaling makes our plain old telephone system a multipurpose information instrument. The CCIS signals are out-of-channel signals not heard by the subscriber and are for managing the network:

• To inform a switching center that for one reason or another the call cannot be completed, or
• To inform the center that the call has been completed and to disconnect and start a new call.
• To manage the flow of messages in the network by keeping track of and determining the status of various transmission paths in order to avoid trouble spots and route calls around congested routes.
• To trouble-shoot the network and isolate trouble spots for dispatching maintenance and repair services, and
• To collect and process billing information for the use of more than 1500 telephone systems in the nation and for international systems as well.

Finally, the common channel signaling system provides the means by which the plain old telephone system of yesterday is learning how to deliver the services that are confusing to the general manager and to which we turn now.

NEW SERVICES FOR NEW REVENUES

It was a great boon to the telephone company when the plain old twisted pair system learned to deliver varieties of new consumer and business services.

Additional uses for the traditional plant without a major upgrading of the system meant additional revenues with relatively little new plant investment. Remember, the message network is still the same analog network that has been in place for more than seventy-five years and where the major telephone plant costs have been "buried"; twisted copper pairs of wires still travel into every subscriber's household, office building, and local switching center, accounting for about 70 percent of the $162-billion telephone plant investment in the nation.

Over the years, the long-distance plant had been upgraded, as we saw in Chapter 6 with the installation of broadband microwave and optical fiber transmission links. And let's not forget the satellite which has become a major distributor of long-distance telephone services. Switching, too, has emerged from the mechanical Strowger technology to the stored program computer-controlled switches we know to be among the largest computers in operation today. The introduction of control signaling, and especially common control signaling systems, has created opportunities for telephone companies to use their existing analog system in ways that could attract new revenues without requiring replacement of a major portion of the plant. Consequently, with a relatively small investment, primarily in semiconductor devices, new revenues are being found from innovative uses of the traditional telephone plant.

Electronic Convenience Services

If you were to remove the cover of your telephone you would find a good deal of empty space, certainly enough for several semiconductor microprocessors. It is not surprising, then, that the telephone should become a convenient place to house those electronic convenience devices we have learned to depend upon—alarm clocks, radios, timers, stopwatches, calculators, and message recorders. Speed dialing of emergency and frequently called numbers also can be put into that general category of electronic conveniences housed in the telephone. Who can guess what more we are likely to find in our telephones in the future?

Much more important, however, are the services delivered through the telephone that provide for communications when we are separated in time as well as space. Modern telecommunications technology has achieved what the ancients and science fiction buffs alike have long searched for—instant transfer across space and time.

The Telecommunications-Transportation Tradeoff

In 1902, H. G. Wells predicted that in the twentieth century the growing use of new means of communications to conduct business at a distance would reduce land transportation. In 1914, *Scientific American* forecast a decline in transit congestion as new communications services grew. When the Arab embargo of 1973 led to an almost tenfold increase in oil prices, it was widely assumed that auto travel would be significantly replaced by the telephone and several researchers showed just how much energy could be saved by trading transportation for telecommunications. But travel and congestion, in the air and on the highways, seemed to grow as rapidly as the new communications technologies arrived on the scene.

The latest estimate of the cost of executive and staff travel by U.S. firms was in the neighborhood of $70 billion per year and rising. Transportation accounts for about 25 percent of the nation's total energy use with automobiles consuming almost 42 percent of this amount. The latest census showed that household budgets for costs connected with commuting to work, shopping trips, entertainment, and visiting are growing more rapidly than other expenditures. If only 10 percent of executive travel in the nation could be reduced—by eliminating two three-person trips per month between New York and Chicago, or between Los Angeles and San Francisco, for example—a savings of perhaps as much as $4 billion in time and travel costs could be realized, taking into account the cost of the telecommunications substitute.

There are many theories as to just why the convenience of electronic communications has not widely turned commuters into "telecommuters." Past experience with audio conferencing systems generally supported these theories. It *was* difficult to set up a conference; you *had* to call the conference operator who in turn called all of the conferees a day or two in advance of the proposed meeting time. When finally "on-line," voices faded in and out at the most inconvenient moments and often seemed to be coming from the bottom of deep wells, where the participants were surely up to their chins in water.

Researchers in Great Britain and Canada, where audio conferencing attracted a great deal of interest in the 1960s and '70s, argued that audio teleconferencing was not a good way to meet people for the first time and that the conference would be more successful if the conferees were well known to one another. They said that telephone conferencing was not suitable for getting to know people, or for bargaining and negotiating. They were greatly concerned with what they called "social presence," a synthesized measure of several factors that affects the quality of human communications: the ability to transmit facial expressions, posture, the direction the individual is looking, dress, and other non-verbal cues. All of these factors contribute, they said, to the social presence of a medium; audio conferencing using the plain old telephone did not rank well when compared to a multimedia system. This should give us clues as to how we might improve audio conferencing.

All was not lost, however, for the researchers also showed that despite its relatively poor social presence telephone conferencing was, nevertheless, quite a good medium for giving orders, making quick decisions in a crisis, settling differences of opinion, offering briefings, and seeking information, just the sort of activities sales managers might engage in with salespersons scattered around the country.

There is growing evidence, however, that as people do more conferencing by telephone the disadvantages these researchers cited disappear. Telephone conferenced meetings tend to be shorter than in-person meetings and those who engage in teleconferencing feel they are more persuasive in their arguments and that their fellow participants are more trustworthy. Despite the early fears that hearing disembodied voices would make conferencing so impersonal as to be uncomfortable, frequent users of audio conferences are becoming quite happy with the results obtained. Furthermore, they now find themselves doing all of the things early researchers said would not work well: bargaining, negotiating, and meeting new people for the first time. With practice, the social presence of the telephone audio conference seems to be improving; as is so often found with the communications technologies, people soon learn to fit the instruments to their own particular needs and habits and both adapt to and appropriately adopt the instrument.

The smart telephone capable of delivering what is becoming known as "custom calling" options has gone a long way toward making telephone conferencing easy to arrange and almost as effective as the in-person meeting. You can set up the conference from your own telephone, control access to it, and encourage participation by making it easy for everyone to come on-line. The silent type, afraid to enter into a discussion around a table, rushes in eagerly by telephone where the push of a button informs the chairperson that he or she wants to make a point.

When telephone conferencing is to be used frequently—in a multilocation insurance business with weekly loan decision meetings or at an airline where emergencies arise that must be handled quickly, for example—a conference control telephone can be installed. This device dials up participants using a card dialer (a punched card is read by a small card reader in the telephone), keeps track of who is in on the conference, and indicates when a participant wishes to speak. Often, during a conference, questions are raised that cannot be answered by those at the meeting. The chairperson simply calls up the "expert" who comes on-line to respond to these questions. Faced with a "room full" of clients seeking advice, a surprised consultant's mettle is soon tested.

Conference "bridges" are located throughout the telephone system. These devices keep track of the amplitudes of the signals from telephones in the conference, comparing them against each other and balancing voices to ensure that all of the participants "are in the same room" acoustically if not actually. Frequent users of teleconferencing will often install their own in-house bridges so that they are assured of access when required. Well-designed and well-adjusted bridges eliminate fading and participants will no longer seem up to their chins in water.

The Visual in Audio-Visual Communications

Visuals also can be introduced to the telephone conference. Remote writing or telewriting using "electronic blackboards" can be dialed up on a separate line and controlled through the conference telephone. Information written with a tiny "location indicator" attached to the writing instrument on special conductive surfaces (the electronic "blackboard") is translated into signals of varying frequencies, a form of frequency modulation. These signals are transmitted, stored, and reproduced as continuous writing at the receiving end.

If participants wish to use slides in their presentations, projectors can be activated at each location by sending coded signals very similar to the kinds of multiple frequency tones used for pushbutton dialing. The remote projectors decode these tones converting the signals into instructions for the projector to step ahead or return to a previous slide. You can flash pictures of the participants as they speak if you think this would make the conference more personal.

In Chapter 2 we found that in order to transmit a given quantity of information you need a specific amount of time and bandwidth. Further, we found that we could tradeoff time and bandwidth in order to send information. Thus, we could send the information on a 78 RPM record 2⅓ times faster than at 33⅓ RPM by increasing the bandwidth or frequency range by 2⅓.

We can do the same with video images. It requires about 4MHz (4,000,000 cycles per second) to send a black and white television picture. But with only 4KHz of bandwidth available on the voice channel we must increase the transmission time. So instead of sending the video in "real time," we record the images at the transmitting end and send one frame every thirty seconds; this is called "freezing" the pictures or frames. The viewer scans the video image at a slower rate than it is being recorded; hence the term "slow-scan" television.

Slow-scan television delivers a relatively sharp but jerky picture, with each frame sliding across the screen in the form of still pictures, much like a slide show. Normally, the slow-scan signal will occupy the entire bandwidth of the voice channel. You will need a second audio channel if you want to talk while you view.

Documents for a Paperless Meeting

In Chapter 2 we found that Morse's telegraph was a very early form of a digital system. Remember how we replaced the dot and dash by a "zero" and a "one" and how this "digital" signal was transmitted along the analog telephone system? The ability of the analog telephone to transmit digital signals at relatively slow speeds, usually in the neighborhood of 300 to 1200 bits per second, without altering the transmission technology makes it possible to send other forms of visual information to the participants in the audio conference. For example, relatively slow-speed transmission of documents and other illustrations can be transmitted as facsimile or FAX messages. Similarly, teletype messages, the very modern version of Morse's telegraph where the dots and dashes of his code are replaced by the letters of the alphabet and numbers, also can be transmitted.

In our next chapter, we shall have more to say about digital transmission along the twisted pair telephone wires and about facsimile, teletype, and the switched version of Morse's telegraph known as the Telex. We also shall discover how the audio conference can be further enhanced by transmitting data from remote computers to the conferees over the plain old analog telephone system.

Audio teleconferencing capabilities have been available pretty much from the beginning of telephone communications. The telephone operator at the switchboard could have arranged for multiple parties to be on the line and often did so without warning or telling the unfortunate, and often lovesick, conversationalists. Even with the disappearance of rows of operators, it was still possible to arrange for teleconferencing, but the promise of "letting your fingers do the walking" did not come to pass. We have explored some of the features of the intelligent telephone that are making the telecommunications transportation tradeoff an economically meaningful reality. What with the constant threat of energy embargos, shortages, rising prices, and air pollution there are certainly many good reasons why society should exploit this potential.

So far we have succeeded in "teleporting or beaming" people across space, very much as "Trekkies" are known to do. Now we shall do what even the "Trekkies" cannot do—transport people across time.

The Telecommunications-Time Tradeoff

We have remarked, more than once, that the telephone is our most modern of telecommunications technologies. Not only is it interactive, ubiquitous, and easy to use, it is immediate. But how often have you initiated a call and been connected only to find that the person you are calling has "just stepped away from his/her desk"? And how often has that person returned your call just when you stepped away from your desk? "Telephone tag" is a plague upon telephone users in almost one out of every five telephone calls! What we need is a means to provide communications between people separated in *time as well as space*.

When the central office consisted of a human operator at a switchboard, we could leave messages for later delivery or instruct the friendly operator to intercept calls when we were busy or away and take messages for us. There are many tales of operators performing these services almost too well; privacy was not often respected! But the telephone *did* leap across time as well as distance.

When the "pleasures" of the human operator disappeared with the arrival of automated switching, we married audio cassette recording techniques to the telephone. We have all suffered the inconveniences of messages incompleted and often entirely lost on scrambled tapes and cassettes. How do we know when we have run out of recording tape, or if the answering machine has failed? Too frequently we discover problems too late—the message is gone, the customer "turned off," the friend alienated because we did not get the message.

The messages we hear (and leave) are frequently performances rather than messages since we have so little tape (memory) with which to work. Despite their desire to create a human communications environment across time and space, answering devices often create one that is quite inhuman.

We cannot locate the necessary banks of answering machines with bulky tape decks likely to be required to handle the message needs of an exchange at the central office and, even if we could, searching through them for the message requested would be excruciatingly painful for the caller and would tie up switches and lines. A mass higher-fidelity voice storage system is therefore required.

Digitizing analog signals permits us to send many messages in the same bandwidth space by means of time division multiplexing (see Figure 2-18). We interlaced these digitized analog signals in order to increase the capacity of the transmission line and thereby achieved considerable transmission economies.

The same technique can be used for storing digitized voice on magnetic disks similar to those used by computers for storing large amounts of information. Just as we sampled the voice signal for transmission we sample the voice signal for interlaced storage on the disk, indeed for storage on many disks in the mass storage facility. This facility is accessed by the electronic switch, either in the telephone central office or in the firm's private automatic branch exchange.

Telephone users want to recognize the person with whom they are talking; neither the computerized Hals of "2001" fame nor talking dolls are suitable for voice store and forward or voice answerback services for which people will pay. Digitized voice must have human qualities.

In Chapter 2 we found that in order to reproduce a signal accurately we must sample the analog signal at twice its maximum bandwidth. A bandwidth in the range of 300 to 3800 hertz, the range of the human voice, delivers signals that are recognizable and, indeed, voice signals that are quite human. We must, therefore, sample this signal 8000 times per second. To ensure that we accurately capture all of the nuances represented by the amplitude of the signal, we divide up the amplitude in a number of discrete levels, as shown in Figure 7-7.

It would require an infinite number of levels to represent all of the nuances of the original signal. Telephone engineers have found that subdividing the amplitude into 128 levels using an eight-bit code delivers very recognizable voice qualities. You will often hear engineers talk about a voice channel capable of sending 64,000 bits per second or sixty-four kilobits (64Kb/s). They arrive at this simply by multiplying the sampling rate (8000 times per second) by the eight bits required to represent the signal.

Voice store and forward services enable us to trade telecommunications for time. Consider the following familiar scenario:

You must be away from your office or home for a time and wish to leave instructions or questions for your family or staff with a request that they provide the information to you through the store and forward message service. You are free to travel about and yet you can still receive desired information and leave necessary instructions. Should you wish to schedule the delivery of your messages, you may record your words for later delivery to those who are to receive your messages. It's like having the friendly village operator back again, but with fewer embarrassing risks.

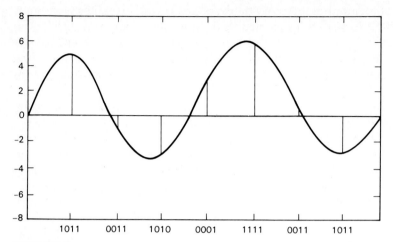
FIGURE 7-7
Quantizing and Coding a Voice Signal (4-bit code).

For the telephone company, this voice store and forward capability creates many new sources of revenue in addition to business and personal message services. In many cities around the nation, there are 900 numbers which can be called in order to receive special messages, such as the latest sports scores and news, weather, jokes, and inspirational messages of all kinds. An "information" charge is generally added to the usual telephone charge for accessing the 900 number.

DISTRIBUTING INTELLIGENCE IN THE TELEPHONE SYSTEM

The telephone operator at a switchboard—plugs and jacks in hand—provided the intelligence the telephone system required in its early years, and, indeed, well into the twentieth century. The operator would dial, forward, and store (remember) messages for later delivery, set up conferences, albeit limited to perhaps two persons, keep track of billing information, be on the lookout for trouble on the system, and call for help when there were problems in the network.

When businesses learned the value and importance of good communications and increased the number of telephones in their offices, they installed a switchboard and an operator at their own facility. This was the first Private Branch Exchange (PBX), and it was as intelligent as the persons who operated it.

When mechanical switching was introduced into the central offices, it was also introduced into the Private Automatic Branch Exchanges (PABX) in offices. These were large and noisy devices hidden behind heavy doors often near to receptionists desks in outer offices. But as computing became an integral part of modern switching, semiconductor microprocessors reduced the PABXs to desk-top units that were small, attractive, and intelligent. These systems were, essentially, the common control mechanisms we discussed in Chapter 6 and illustrated in Figure 6-5 and 6-6, but without the large switch matrices found in the PSTN switching centers.

Today's telephone system is a dual communications network; one network performs the task for which the telephone was invented—the transmission of people-to-people messages—and a second network controls, manages, and ensures that the people-to-people network works efficiently. Control signaling travels along this second network, interconnecting seats of intelligence along it. At these locations we collect the information necessary for network management and control, store this important information, and use it in ways that enhance the services the message network can deliver. This intelligence is distributed at the various switching centers and at PABXs located in offices. Today, intelligence is also located in the telephone instrument itself.

The general manager's telephone is an intelligent device containing many of the functions found both at the switching centers and in PABXs. So, it seems, intelligence is finding its way back to where it originally began, with the user of the telephone who rang up the local operator and told her (it was almost always a her!) what number to call, with whom he or she wished to talk, what to do if that person was not in, or where the person might be found, to connect in another person, to record a message on her desk pad and pass it on when the intended receiver could be found and called, to keep track of the bills, and more.

THE POWER OF VOX HUMANA

A major office automation and computer manufacturer recently decided to eliminate "telephone tag" in their executive and R&D offices by introducing computer mail: people "talking" to people by way of their computer terminals. With messages stored, forwarded, and waiting for the recipients, there would be no more busy signals, no more missed messages, and no more "telephone tag". Three weeks after the introduction of electronic mail services, the computers were overloaded; it seemed that everyone quickly took to this voiceless messaging system despite early warnings that the old-timers would not accept so radical an innovation. Management now faced the problem of reducing the overload on the computers and hit upon the idea of charging for electronic mail as a way of eliminating unnecessary personal non-business messages.

In the process of determining a reasonable tariff, a voice mail system was introduced on a trial basis. Even before the new tariffs could be introduced, computer mail demand had all but disappeared as everyone shifted to voice mail for everything except long document transfers. The human voice conquers all!

No surprise! After all the human race has invested several thousand years in the development of language and has become rather accustomed to the sound of the human voice. Despite the often high cost of meeting people face-to-face it is rare that we humans would willingly substitute some other means of communicating. The telephone has, however, provided us with a wonderfully convenient and increasingly flexible means for communicating when a face-to-face meeting is either physically impossible or emotionally undesirable.

The telephone system is one of man's most remarkable achievements. It is extremely complex, yet easy to use. It is larger than life yet it presents a human face to everyone. It has affected society in ways that have confounded expectations. It is more than 100 years old but it continues to meet the needs of an increasingly complex society. It appears forever new and forever young.

Almost 70 percent of the capital investment of the telephone system has changed only slightly over the past fifty years; the twisted pair copper wire on poles and in underground ducts are still in place. When was the last time you replaced the telephone wiring in your house or apartment? Except for exchanging your rotary dial model for a pushbutton one and, perhaps, for a different color, your telephone has not changed appreciably. Yet, plain old telephone services are no longer plain and can hardly be said to be old. Innovations in the telephone system have been brought about by investments in a relatively small portion of the system, almost always invisible to the consumer. The genius of the telephone system is that it is readily adaptable to change, to becoming more intelligent. We have seen how computer technology, memory, and microprocessors, have increased the intelligence of this system. We have seen how this intelligence is providing a system suitable to our changing behaviors and needs. What more can we ask of high technology systems other than that they respond to human needs and adapt to human behavior? Isn't this what we mean when we refer to a technology as being appropriate? Certainly the telephone is an appropriate technology.

Not quite, you say! What about transmitting high-speed data, good-quality pictures and graphs rapidly? What about real television, not the pseudo-television of slow-scan slide shows? What about meeting our need to communicate while on the move—from an automobile, airplane, ship, or train?

In the next chapter (and in Chapter 10) we shall explore these "deficiencies" and examine just how telephone systems throughout the world are working to overcome them.

ADDITIONAL READINGS

If you ask an engineer to explain the technology of intelligent telephones you will probably receive a rather quizzical response. "Not much different from any other telephone," the engineer is likely to say. And that is probably a sensible answer. The beauty of the design of intelligent telephones is that they are, indeed, modernized plain old telephones with judicious use of microprocessors. The wonder is that with so little change they can achieve so much.

The literature on the technology of intelligent telephones is rather sparse, but there are a number of excellent writings on how to make the best use of these devices.

The Bell System Technical Journal, "A Voice Storage System," May/June 1982, Volume 61, No. 5, pp. 811–911. The engineer's view of one aspect of the intelligent telephone.

Johansen, Robert, Jacques Vallee, and Kathleen Spangle, *Electronic Meetings: Technical Alternatives and Social Choices* (Reading, MA: Addison-Wesley, 1979). A comprehensive analysis of teleconferencing concentrating on management and other societal issues.

Martin, James, *Future Developments in Telecommunications*, 2nd edition (Englewood Cliffs, NJ: Prentice-Hall, 1977), pp. 27–131. Once again, a valuable reference.

Meadow, Charles T., and Albert S. Tedesco, eds., *Telecommunications for Management* (New York: McGraw-Hill, 1985), pp. 213–226.

Nilles, Jack M., F. Roy Carlson, Paul Gray, and Gerhard Hanneman, *The Telecommunications-Transportation Tradeoff: Options for Tomorrow* (New York: John Wiley & Sons, Inc., 1976). An early and somewhat enthusiastic view of the prospects for the tradeoff.

Reid, A. A. L., "Comparing Telephone with Face-to-Face Contact" in *The Social Impacts of the Telephone,* Ithiel de Sola Pool, ed. (Cambridge, MA: MIT Press, 1977).

Short, John, Ederyn Williams, and Bruce Christie, *The Social Psychology of Telecommunications* (New York: John Wiley & Sons, Inc., 1976). One of the very few studies of the social-psychological aspects of the telephone with some early conclusions that are now being challenged.

CHAPTER **8**

BITS, BYTES, AND BAUDS OVER THE TELEPHONE: HOW TO TALK TO YOUR COMPUTER BY TELEPHONE

HUMAN COMMUNICATION IS A MULTIMEDIA EVENT

Many of us have pored over a Radio Shack or L.L. Bean catalog, found just what we wanted, and placed our order by telephone. But have any of us ever stopped to examine the many different types of information transmitted and received throughout that transaction? Let's look at it in detail and see just what goes into shopping by telephone.

We have just finished scanning our latest computer catalog and the spirit moves us, right then and there, to purchase that computer game we've been considering. We call the number given, the order taker gets on the line, and prompts us to begin the transaction. We say that we want to buy the new XAXXON game (Wow!, we can "dodge missiles, blow up fuel tanks, zap the enemy fleet, and match our wits with the deadly XAXXON Robot himself! 32K required" number 28-4052) in the latest catalog received. There is a pause while something goes on at the other end and, finally, the order taker returns, presumably having checked that there is, indeed, a XAXXON game in the catalog. The very efficient order taker reads back our order and we confirm that yes, XAXXON, number 28-4052 is exactly what we want.

"How do you wish to pay for this order?" we are asked. We hunt for a credit card and read off sixteen numbers, and three additional ones in response to the usual request for card expiration date. There is another long pause, indeed, a very long pause, while our credit is checked.

Happily, the clerk returns and now wishes to know if we intend to pick up the formidable XAXXON ourselves or have it mailed to us. We decide to have it mailed and provide our address: four house numbers, the street name which we are asked to spell (in this case fourteen letters), the city (another sixteen letters), the state, and finally the five digit zip code. For good measure the clerk asks for our telephone number, just in case.... We hope there is no "just in case...."

This most efficient clerk now repeats the details of the entire transaction—numbers, words, descriptions, addresses, and all. At long last we hang up the telephone and wonder why in the world we ever thought that purchasing by telephone was such a great way to do business.

Meanwhile, our order taker has called the warehouse only to discover that the particular model we asked for is unavailable but on order and that another version using cassette rather than disk is available. We are on the phone again, this time pondering how in the world we could possibly use a cassette version when the computer on which the game is to run does not have a cassette drive. Should we start the whole laborious transaction over again and order the "Ogres of Orion" so that we can savor "each victory over an awesome beast that will get us closer to our ultimate opponent—the evil wizard," or quit the entire business and visit the store in person?

Human communication does not depend entirely on the sound of the human voice; eye contact and body language are extremely useful for transmitting information. Have you ever carefully observed someone talking on the telephone—for example, a salesperson making a sale or an executive making some earth-shaking proposal? You will see a great deal of animated arm swinging as the speaker paints a verbal image of the sale or proposal being made. When the conversation reaches a critical stage the speaker will call on an assistant for the cost estimates, pro formas and budgets, and the arm swinging will be replaced by furious scribbling and pencil pointing as the speaker tries to deliver numbers over the telephone.

Human communication is a multimedia event. The human voice is quite good at performing multimedia functions over the telephone, painting word pictures and delivering "data." Some voice qualities are better than others; that's why we often describe a person as having a powerful telephone personality. But if we actually could have sent and received data and pictures over the telephone wires, the purchase of our video game would probably have been completed in half the time and with considerably less frustration.

DEMANDS FOR COMMUNICATIONS DIVERSITY ON THE CARRIERS

Today's business activities whether carried out from home or office, require the simultaneous use of many media. The competition and complexity of today's working environments make it desirable that communicators be able to use voice, data, and one or more forms of image distribution, full- or slow-motion video, or facsimile. This need for diversity in communications emerged in the United States in the mid-1960s. The post–World War II industrial and business explosion had resulted in the distribution of plants and offices throughout almost every corner of the nation and leaped across oceans as national firms went multinational. Communications and transportation costs grew rapidly, and, before long, computers wanted to talk to one another even more frequently and for longer periods of time than did the human workers in plants and offices.

It was during this period that pressures increased on the telephone carriers to provide more diverse forms of communication services at rates more fairly reflecting their actual costs. The high costs of corporate telecommunications and a growing demand for data and visual communications encouraged the telephone carriers to find ways to lower costs and deliver more varied communications services on the telephone without having to completely replace their multibillion dollar capital investments.

In the previous chapter we found that it is possible to send and receive some pictures by telephone using a movable slide show called slow-scan television. We also can deliver sketches and figures via the electronic blackboard. But neither of these tools is generally available in most firms, to say nothing of homes and small business offices, and even when one is, it can be rather expensive, slow, and burdensome to use. Today, a considerable number of us communicate with personal computers, accessing remote databases and sending computer mail to our networked friends. In many branch offices of multi-plant firms, word processors and facsimile devices are interconnected by the telephone. So the telephone *can* do more than talk!

In this chapter we explore just how the telephone, designed for the voice and used to carry it for more than 100 years, can respond to this demand for diversity.

DATA ON THE VOICE NETWORK

The telephone is our primary means for the exchange of information. It is designed for voice transmission. Speaking over the telephone is a continuous operation; we may stop to think or wait for our party to speak, but these brief silences between our words are all very important for capturing the personality of the speaker. Once on the line, the connection is held until the conversation or silences are completed and we ring off. Despite what comedy writers may have led us to believe, most telephone conversations last on the order of three to four minutes. The telephone network was optimized around these basic characteristics of human communications.

Methods have been incorporated to extend the transmission distances over the local loop by adding relatively simple devices known as loading coils at regular intervals along the line. These coils tune the line, much as we do our radios when we tune to a station at a selected frequency. In the radio tuning process, we allow only the frequency of a particular station to be delivered to us. In the same way, loading coils allow voice frequencies (in the range of about 300 to about 3400 hertz) to be transmitted with less attenuation than other frequencies. By way of these loading coils we are able to extend the range of the twisted pairs for voice transmission; indeed, the first long-distance telephone call between Boston and San Francisco was delivered over open wire lines using loading coils.

While we transmit voice over the local loops with the aid of loading coils, higher frequency signals are faced with higher levels of attenuation or losses. Remember our discussion of bandwidth in Chapter 2? We noted that a pulse, such as a dash in a telegraph signal (shown in Figure 2-12), required a very large bandwidth. The more pulses we transmit in the same period of time, the higher the frequencies the transmission line will have to carry. Clearly, a transmission line that has been designed for voice frequencies in the 300 to 3400 hertz range will not perform as well with a series of fast pulses such as data in the form of bits.

Now recall our discussion of the telegraph. We found that its dots and dashes were really digital signals. Two symbols of different time values are assembled in various ways to represent the letters of the alphabet and the numbers from one to nine. We substituted a zero for a "dot," thereby creating a Morse Code consisting of "ones" and "zeros."

Early telegraph systems were operated on open wires strung between telephone poles along highways or railroad tracks (rights-of-way), and used a continuous direct current which was interrupted to send a dot or a zero as shown in Figure 8-1.

If we send these pulses along telephone cables there is a great deal of distortion. The 500 microsecond pulse we transmit in Figure 8-2 appears much wider when it is received, in fact, almost three times wider than what was sent. This variation in width causes a great deal of interference between pulses so that we do not know when a series of digital signals has begun or ended, let alone if the signals received are zeros or ones.

While it is possible to send pulses in this manner on the subscriber loop, perhaps over distances as long as ten miles, it is not possible to send such pulses along the trunks between switching centers. Because many voice signals must be frequency multiplexed along these trunks in order to minimize the need for expensive copper and duct space, these trunks do not carry direct current; they carry high frequency carrier signals. The voice signals are assigned "chunks" of frequencies or pass bands of about 300 to 3400 hertz. These pass bands are

FIGURE 8-1
Sending Morse Code by Interrupting a Direct Current.

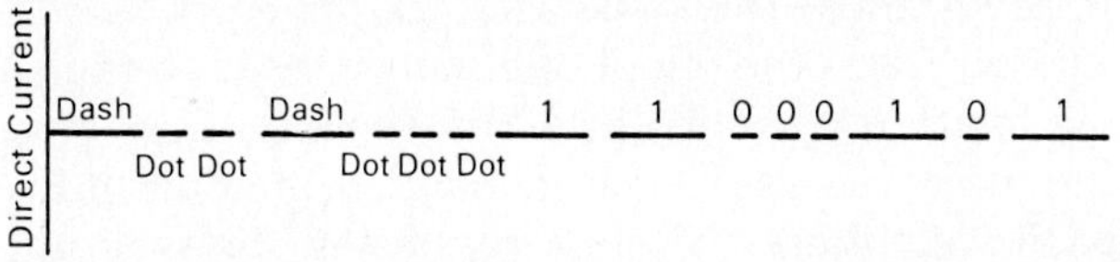

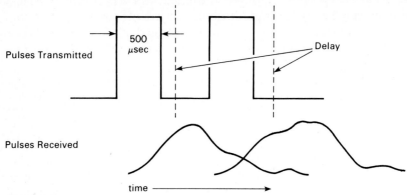

FIGURE 8-2
Pulses Running into One Another Creates an Interpulse Distortion.

used to transmit our pulses, not as simple on/off signals as shown in Figure 8-1 but as modulated audio tones, or tone signals, between 300 and 3400 hertz similar to those used to send Morse Code (see Chapter 2). To send pulses or digital on/off signals along a telephone line as tone signals requires devices that can translate the pulses into modulations of the carrier at the transmitting end and that can demodulate the resulting modulated carrier signals into pulses or digital information at the receiving end. These devices are the *modems* (short for *mod*ulator-*dem*odulator) for sending the data from your personal computer on a telephone line.

We have encountered the transmission of digital signals by means of tone signaling several times in our study of modern telecommunications. In Chapter 2 we described Morse Code on a carrier as a form of on/off tone modulations we called on/off keying (OOK) (see Figure 2-5(b)). And in Chapter 6 we found that the signals required to set up a telephone call required tone signaling to send dialing pulses, as shown in Figure 6-2. Called number information is transmitted between exchanges by means of tone signaling using the standard frequency of 2600Hz at a rate of eight to ten pulses per second. The clever use of six tones at different frequencies allows the telephone system to represent the digits 0–9.

The traditional telephone, our POTS, without the addition of many modern computerized technological "fixes" can transmit digital signals, that is, on/off signals, at suprisingly high speeds using tone-modulated carriers. Over the years, the telephone system has performed quite well delivering signals between teleprinters or automatic telegraphs, the most familiar of which is the Teletype ™. While in the United States the telephone outdistanced the teleprinter, in Europe switched teleprinter networks called Telex systems were quick to be adopted and for many years served as the basic communications network in the absence of the telephone for most businesses. We shall use these automatic telegraph or teleprinter networks to describe how data can be transmitted by telephone and, in this way, begin our education into the mysteries of computers and data communications.

Bits, Bytes, and Bauds

Most teleprinters transmit between sixty and 100 words per minute. An average English word consists of five letters and one space, or six characters. These characters, of course, must be translated into bits and a five-bit code known as a Baudot code has become standard for most teleprinters throughout the world. (The French telegraph system adopted Baudot's five-bit code as early as 1877.) Each letter of the word, then, consists of five bits (two to the fifth power; $2 \times 2 \times 2 \times 2 \times 2$ is equal to 32, which is just right for the 26 letters of the alphabet, and for any six punctuations). In order to know when a word begins and ends we must add two additional bits. Except for the bit that informs us of the end of the word, all of the bits are equal in time; for purposes of this example we shall use the European standard with the stop bit being half again as long as the other bits. For each character of our word, therefore, we transmit 7.5 pulses (1 + 5 + 1.5 pulses per character). A group of bits to be sent or processed together is called a *byte*. As we shall see in a moment, this coding is widely used to establish the protocols for asynchronous communications, the kind of communications in which computers are engaged.

A teleprinter system operating on a telephone line sends fifty pulses per second. A pulse per second is called a *baud*, again in honor of the Frenchman, Baudot. At fifty baud the teleprinter transmits about sixty-seven words per minute. At this rate, we can multiplex many such teleprinters on a voice channel. Because there is the "spill over" or interpulse distortion shown in Figure 8-2, the number of digital transmissions on a voice channel is usually limited to twenty. In a single voice channel of between 300 and 3400 hertz, digital transmissions are assigned bandwidths according to the speed with which the data is transmitted. If a transmission is to take place at 110 baud, 170 hertz is assigned, for 220 baud, 340 hertz, and for 440 baud, 680 hertz. Note that there is a very approximate correspondence between a baud and a hertz (a useful rule of thumb to keep in mind in our upcoming discussion of transmitting digital signals on telephone lines).

Protocols: The Rules of the Road

When speaking over a telephone we can easily recognize words and sentences and readily know when they begin and end. When we communicate through a satellite link, the quarter second delay from transmission to reception is often disconcerting and we are tempted to say "over" or "stop" after every sentence. Indeed, radio operators communicating on *simplex* lines—that is one transmission path for both speakers—do use "over" and the familiar "over and out" to make their communications efficient and intelligent. These are the rules of the road, or protocols, required for communicating in special situations.

Communicating information as sequences of ones and zeros is one such special situation for which we need rules of the road to know when a "word" has begun or ended, or when a message has been completed. A series of decision rules have to be built into the sending and receiving devices to permit one to understand what the other is doing. For example, suppose we sent an 8-bit per second stream of bits like this: 0 0 1 1 1 0 1 0. If the receiver thinks that the transmission rate is four bits per second, rather than eight bits per second, the word will be read as 0 1 1 1.

The receiver must be *synchronized* with the transmitter so that they work in step with one another. In both the receiving and transmitting equipment, devices generate accurate internal pulses that act as clocks; indeed, they are often called *event clocks* since they are turned on when a transmission event is begun. When a string of bits representing a message is transmitted, a certain number of bits is used as a code to tell the receiver that a message is coming, that the clock is running, and the receiver begins to "read" the message.

Suppose that the following data stream is received and the receiver only knows that it is the ASCII 8-bit code (American Standard Code for Information Interchange, the standard for digital communications over telephone lines):

10110100011011100110001101001110010100101011, etc.

What are the boundaries of this group? Where does the word begin and end? ASCII has one code for this purpose called "synch" and this code is:

011010011

The receiver will look for this code in the bit stream. When it finds the code, the receiver will count the remainder of the stream in groups of eight, since it knows the information is being received according to the ASCII standard. Try it with the long stream of numbers above. Unless the printer has made a mistake, you will find that the message is:

101   END (of previous message) "synch" START (of new message)

The receiver will not be able to decode the first three bits and ask for a repeat of the synch in order to make certain that these three bits are, indeed, noise. You are likely to hear more about this *framing* process in your travels through the data communications business.

Framing is but one level of what are usually many levels of rules. When the clock starts running, a long burst of repetitive bits is sent, such as a string of zeros followed by a string of ones to which the receiving clock adjusts itself. Then there is a string of bits that tells us that the transmission of the message has ended. In *synchronous transmission*, "start" and "stop" signals are not used; one string of information bits follows another until there is a recognition that the "sentence" is completed and the event is over.

Computers generally communicate *asynchronously;* indeed most Telex and data communications are asynchronous communications. In asynchronous communications the message is broken up into short groups corresponding to the code for one character. Thus, the Telex Baudot message is a 5-bit group;

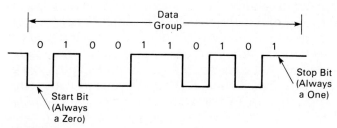

FIGURE 8-3
Protocol for Asynchronous Data Communications.

most personal computers use an 8- or 16-bit group and some new ones are opting for a 32-bit group. (This is what the manufacturers talk about when they vie for the market with their 16- or 32-bit machines; these machines process and transmit information using words of either 16- or 32-bit lengths. The larger the word they can process, the faster the machine.)

In asynchronized communication, there is no event clock, so synchronization between the transmitter and the receiver must be reestablished for each group. This is accomplished by adding extra events such as a start and a stop bit as shown in Figure 8-3. Eliminating the need for event clocks in modems reduced their cost; hence the use of asynchronous communications for most data communications networks using the telephone system.

If you use a personal computer (or have seen a friend use one) for communicating with other computers or "logging" into public information services such as The Source^{sm}, you must provide the receiver with instructions as to your particular "rules of the road." First, you must tell the receiver that you will be transmitting at a certain rate—300 or 1200 baud—then you inform the receiver that each word is a particular number of bits in length, say seven bits, and whether there will be one or two stop bits. One other important rule must be provided. How does the receiver know if it has received the correct word, the correct number of bits? It is entirely possible, indeed probable, that a bit could be lost in the transmission process because of excessive "interpulse interference" or noise on the transmission line. This is taken care of by a *parity check*. A redundant bit is added to the bit group or stream of information bits to maintain either an odd or even number of ones in the group. If the receiver detects an odd number of ones when the number should be even, or vice versa, it knows there is an error and that an incorrect number of bits has been received. When that occurs, the transmitter is instructed to send the stream or word again. The transmitter must tell the receiver that it is using either even or odd parity.

Sharing the Expensive Channels

Up to this point in our discussion we used the entire voice channel to send data. This is not very efficient since the transmission of data is "bursty" and there is a great deal of time between bursts of data even in a string that represents a single word. While sending these bursty or discontinuous streams of data we tie up the line and the switch making our connection. Designers assume that the average line and switch are utilized only about 5–10 percent of the time even during the busiest hours. Only when there is a national disaster such as the assassination of a president or a major upset in the Superbowl, and, of course, on Mother's Day, are the lines and switches so tied up you cannot place a call. When computers talk to one another as when the computer "hackers" are at work the lines are likely to be engaged for hours on end. Just imagine what would happen if every computer game buff in your neighborhood decided to play electronic chess with game buffs in other neighborhoods.

The amount of data on the voice network is increasing, not only because of computer hackers but also because more and more uses are being found for data transmission. We are well aware of the increased use of computers in industry, business, government, and education. But there are also the often forgotten slow-speed data transmission required for monitoring fire alarm and security systems in firms and households, and monitoring and controlling energy consumption.

These signal bursts may only travel at 100 baud, or at most 1000 baud, much less than the 9600 baud engineers have found to be the practical limit for a 4000 hertz voice channel connected directly from transmitter to receiver, without any switching in between or with leased private lines.

One way to economize in the use of voice channels is to send both data and voice *at the same time* on the channel. This is done by using only a portion of the channel for voice, say, up to 2040 hertz. This leaves the bandwidth from 2040Hz to 3400Hz available for data transmission. Some telephone carriers have found that they can transmit up to about 960 baud as *data over voice* in that space. At the central office serving the subscriber, the data is separated from the voice signals, or stripped off as the engineers say, and handled from that point onward along separate networks, as sketched in Figure 8-4.

The system serves the subscriber in two very different ways. One circuit allows the subscriber to use both voice and data up to about 1.2Kbs, which is quite suitable for sending computer messages, banking and shopping on the network, and using other services we will discuss in considerable detail later. On the other line, a subscriber can use higher speed business data services up to about 4.8Kb as well as voice. In this way the local loop is essentially untouched; the additional equipment required is at subscriber's home or office and in the central office. The cost of the central office equipment is shared by all subscribers, those using voice only as well as those using voice and data. The additional subscriber equipment is paid for by the subscribers who order these new services.

We noted previously that demands for more diverse telephone services grew rapidly in the 1960s as business firms expanded throughout the nation and, indeed, the entire world. This was also the period when we learned that those soothsayers who predicted that no more than sixty large-scale computers would be necessary to serve information-hungry executives, bureaucrats, and

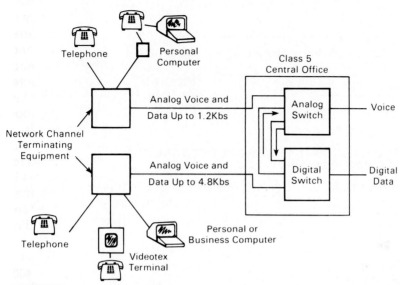

FIGURE 8-4
Dual Services via the Local Area Data Transport Network.

educators were dead wrong. Computer system development moved toward much higher operating speeds and, in turn, created the need for higher data transmission speeds—after all, what is the value of data locked in a computer when it is needed elsewhere, perhaps down the street, across town, or country? Data transmission speeds in the millions of bits per second, or megabauds, were needed.

Responding to these pressures from expanding and geographically roaming businesses the telephone carriers explored innovative ways to meet new demands for data communications without having to redesign the entire telephone network. Many of these new developments, made possible by the introduction of the computer into major telecommunications systems, were economically viable in the line-haul or long-distance service area where revenues were more closely related to the value of services delivered, or, to put it more simply, where service providers could charge a higher price to business organizations.

One of these services was *data under voice,* a means by which very highspeed data, in the range of 1.5Mbs, is delivered under the master group frequency as shown in Figure 8-5. Special coding techniques are used to squeeze 1.5Mbs into a 500kHz space. On the long-distance network, both voice and data are transmitted via the same path; on the local loop, from the toll or tandem switch levels to the central offices, the voice and data networks are entirely separate from one another.

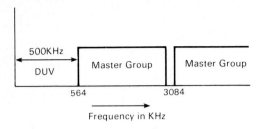

FIGURE 8-5
Data Under Voice on the Long-Distance Carrier.

TOWARD THE ALL-DIGITAL TELEPHONE NETWORK FOR THE TWENTY-FIRST CENTURY

In the previous chapter we discussed how to sample and quantize or quantify the voice signal for digital transmission. In order to capture all of the nuances of the human voice we must sample the signal at twice its bandwidth, or 800 times per second (in accordance with what Shannon told us in Chapter 2), and quantize the signal using an 8-bit code. Thus if we want to send voice over the telephone we must transmit 64,000 bits per second (eight bits × 8000 samples per second).

Time-Space Switching

Imagine yourself at a rather crowded and noisy cocktail party. You hear bits and pieces of several conversations, but you can tell them apart by the quality of the voices. Every now and then a fragment sounds interesting and you "tune in" for a short time only to lose it again when someone talks to you or some other conversation momentarily captures your attention. But you go back to the conversation you were tuned into and pick up some more fragments. You find that you can piece them together and make sense out of the entire conversation even though you missed short segments of it. Sometimes you may not miss these segments at all; the speaker is also interrupted by other conversations, so, without much difficulty, you can assemble the entire sentence or phrase that initially attracted your attention. While you do this, you are probably also "listening in" to other conversations in addition to conducting your own.

You switch your attention from one conversation to another and sample these conversations. You switch in time and space in a random fashion, a process that works only if a very intelligent controller is performing the operation. At the cocktail party you are the intelligent controller. Now let's see how an intelligent switch accomplishes the same task in the modern telephone system.

If we were to time division multiplex several signals over the channel, we could send many signals in the same space (refer back to Figure 2-20). The conversations these signals carry are not meant for the same person. Consequently, we must switch the signals correctly so that samples from the

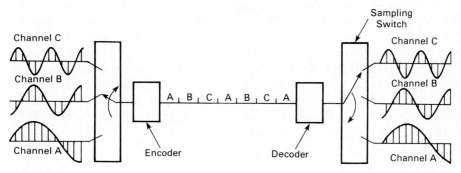

FIGURE 8-6
Time-Space Switching.

many conversations arrive at the appropriate receivers in the proper order so they can be understood. This is achieved by very smart switches that can switch in *time and space*.

Figure 8-6 shows three separate voice conversations, A, B, and C, which are sampled 8000 times per second, coded with the 8-bit number, and ready for transmission. The sampling system takes one sample from A and starts it down the transmission line, then one from B and sends it down the line, and then one from C and sends it down the transmission line. At the other end the signals are decoded and the sampling is repeated. The messages are reconstructed just as you did at the cocktail party, except there are no missing segments. Now suppose that instead of delivering all of the A samples to Channel A we deliver them to Channel B. We have switched the conversation from the person on Channel A to the receiving person on Channel B. All we would have to do is store the samples from A until the sampling switch reaches Channel B and then let them loose.

Since we have sampled in time and switched in space, from one channel to another, this is called *time-space switching*.

The Digital Transmission Hierarchy

The ability to digitize voice and transmit digital voice became feasible in the late 1950s, at just about the same time economic, reliable, and operational semiconductors became available. In response to pressure from business for high-speed data transmission, the Bell System installed the first commercial digital transmission system, the T1 carrier, using both microwave and cable technologies. A family of T-carrier systems has since been offered for line-haul transmission services.

There are time division multiplexed systems that deliver a variety of digital transmission rates, depending on the combinations of carriers used. How many bits per second can be delivered along these transmission paths?

Twenty-four two-way voice conversations can be transmitted along two pairs of wires, one in each direction. With twenty-four conversations at 64,000 bits per second per conversation we are able to transmit 1.5 million bits per second, or megabauds (1.5Mbs). This is the capacity of the T1 carrier system.

Just as there is a hierarchy for frequency multiplexing telephone trunks (see Figure 7-7) there is also a hierarchy for time division multiplexing telephone trunks. For example, we can combine two T1 carriers to provide forty-eight digitized voice circuits and get a data transmission rate of 3.152Mbps (T1C carrier), or ninety-six digitized voice circuits at a rate of 6.312Mbps (T2 carrier), and up to 4032 digitized voice circuits for a rate of 274.176Mbps (T4 carrier).

End-to-end Digital Pathways—the ISDN

Today's telephone system is essentially a dual system, it operates both in analog and digital fashion. Messages are transmitted primarily in analog fashion throughout the system, while control or supervisory signals are transmitted digitally both in the message system as well as along a separate pathway. We have discussed the various ways in which the telephone system has been modernized by the overlay of computer techniques on traditional technologies in order to meet "high tech" business demands while still providing "universal service at affordable costs" for residential subscribers.

This tightwire walking trick has been achieved by modernizing those portions of the system that are used in common by all users, the network's switching and line-haul or long-distance systems. Indeed, without these developments, intelligent telephones with the various revenue-producing services we described in Chapter 7, would be impossible. For today's business users, however, the ability to access a variety of media, voice, image, video, and data at the flick of a switch on an office terminal is becoming more attractive. For the household that is rapidly entering an era where daily chores and transactions are increasingly those of transferring information, having the ability to call up the communications mode most suitable for what you are doing could be very attractive. Remember how our computer game shopping spree turned into a minor disaster? For such transactions to operate efficiently, it is necessary to have an end-to-end digital pathway or transmission system, from office to office, home to home, or home to office, in which video images, data, and still pictures are all transmitted as bits, indistinguishable from one another except in the information they encode.

Is it necessary to have a dual telephone system, one for analog transmissions and one for data transmissions? Not necessarily. We have seen how the T carrier systems handle both for long-distance transmission. Why not a single end-to-end digital pathway?

There are many good reasons for having an "end-to-end" digital system. You can send more messages along a single channel using time division multiplexing rather than frequency division multiplexing. You can integrate switching, control, and messages along a single path thereby monitoring performance more easily, accurately, and reliably. As noted in Chapter 2, signal-to-noise ratios of digital transmission systems are better than those for analog systems; there is likely to be less interference from external high frequency sources. And in any case, it is not necessary to amplify a complex wave form as in an analog system. All we need do is regenerate the pulse in order to amplify the signal without regenerating the noise.

However, there are costs involved. Voice is, by far, the heaviest traffic on telephone systems today, and will still account for more than 90 percent of all traffic on the U.S. network even in the year 2000.[1] There is every reason to expect that voice traffic will also dominate telephone networks in other countries.[2] Humans prefer voice communications and if they cannot talk face-to-face they will substitute the telephone quite readily in almost all cultures when given the opportunity. Young people in France, to the consternation of their elders, use the telephone almost as often as young people in the United States. And even in the Far East, when telephones are made widely available, people quickly adapt them to their particular needs.

To transmit voice on an *all-digital* network requires that you convert the analog voice signal at the sender to bits and back to analog at the receiver. This requires analog-to-digital and digital-to-analog conversion devices at every telephone in the country. With almost 300 million telephones in the United States, all designed for analog transmission, the conversion cost would be astronomical. Where telephones are just now being provided in large numbers the three-chip digital telephone shown if Figure 3-10 can be provided at very low cost driven by economies of scale.

Indeed for countries that are just now installing their telephone infrastructure or greatly expanding an existing one, the end-to-end digital system is viewed as the preferred route for the future. Thailand and Brazil, for example, are just now constructing their telephone infrastructures and have chosen to go "all digital." France is greatly expanding its telephone infrastructure, from about 20 percent penetration a decade or so ago to 80 percent penetration by the end of this century, with an all-digital system. And in the United States our long-distance portion of the telephone networks is rapidly becoming digital.

In new telephone nations, a unique telephone infrastructure is emerging—an all-digital, end-to-end system known as the *integrated services digital network* or *ISDN*.

This network has the capacity to provide voice, video, image, and data services and to provide transmission rates from ten baud for alarm services to 96 megabauds for full-motion television. A subscriber can order whatever service (and transmission speed) he or she desires from the network terminal as shown in Figure 8-7. The ISDN is intelligent in and of itself so there is no need for intelligent telephones and terminals. This characteristic brings up the politics of the telephone.

The Politics of the Telephone

The United States has chosen a competitive telephone infrastructure, one that provides the business and residential subscribers a rich choice of terminal

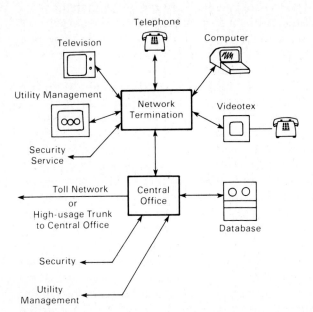

FIGURE 8-7
What the Subscribers Can Order Up on the ISDN.

devices, or instruments, and services. With the divestiture of AT&T the U.S. has expanded an already diversified telephone carrier industry; the seven holding companies in which the former Bell Operating Companies are now housed are independent of one another and their parent. They join the almost 1500 independent telephone companies that have been operated for many years; however, these "new" entities continue to deliver telephone services to the vast majority of households and businesses in the United States. The U.S. has strayed quite far from the monolithic system characterized by the ISDN and conceived by the national postal telephone and telegraph monopolies in many nations throughout the world.

The United States has opted for a marketplace of networks delivering a variety of services to the clients who are willing to pay for them. They range from the plain old telephone services most residences require (now often referred to as *basic* services) to the special services required by business and industry, including digital transmissions, facsimile, teleconferencing, and voice store and forward. The latter category encompasses those *enhanced* services made possible by intelligent telephones and networks and the ability to send bits and bytes over the telephone network.

As noted in Chapter 6 on the telephone, the U.S. prides itself on having universal telephone service at affordable costs; in other words, anyone who wishes to have a telephone can obtain one at a reasonable cost. The carriers have always argued that this would not be possible without cross-subsidy—higher revenues derived from business services support the lower revenues required to achieve universality.

But after divestiture, cross-subsidy is no longer possible. Business and industry subscribers now wish to pay only for the enhanced services they use; they no longer consider it fair to support what many feel is a socially desirable subsidy for low-cost universal residential services. And, it is argued, residential subscribers should pay the "real" cost of their basic telephone services.

Unlike countries that have nationalized postal telephone and telegraph monopolies and where tax revenues underwrite system developments, costs for advanced developments in our telephone system eventually come out of the subscribers' pockets. Consequently we should ask ourselves if it is necessary for all subscribers to have access to the full capacity of services potentially available through an ISDN. Should a subscriber pay for 6.312MBps capacity when he or she only uses 64KBps for voice transmission? Cable television systems provide wired video in the United States; why should a telephone subscriber pay for 96MBps to make video available over the ISDN? Even for the personal computer buffs wishing to access remote databases, 1200 baud is more than adequate, so why pay for 1.544MBps? If the "hacker" wishes to hack at higher speeds, wouldn't 4800 baud over the POTS perform just as well?

We can expect that the specialized carriers in the emerging marketplace of networks in the United States will offer enhanced services, for their business customers and, perhaps, to those households with sophisticated personal computer users. But it is not likely that there will be a single monolithic ISDN in the United States with the architecture envisioned by several European nations and by Japan, a structure as neat as that shown in Figure 8-8(a).

The architecture of the European and Japanese ISDN is shown in Figure 8-8(a). It is quite different from the architecture that is emerging in the United States, shown in Figure 8-8(b). In the European and Japanese systems, the integrated transport network, the all-digital transmission network, reaches from household to household and from business establishment to business establishment. Each and every node on the network has available to it the bandwidth necessary for delivering the services shown in Figure 8-7, bandwidth from ten bits per second to 96 million bits per second.

On the other hand, the U.S. model of the ISDN, shown in Figure 8-8(b), allows for the distinction between basic and enhanced services. There are three transport paths: circuit-switched analog for voice, packet-switched for enhanced data services, and the all-digital system which extends from central office to central office instead of from subscriber to subscriber. While we can expect that, in time, the subscriber loops also will be fully digitized, a competitive market for terminals and enhanced services will likely remain an important factor in the United States.

The decision of the United States to open its telephone system to competition is one that has created considerable concern throughout the world. For as long as most of us can remember, the telephone has been an international instrument; we can call almost everywhere in the world. Over the last decade or so, direct dialing has made it easy to place calls from New York to Cape Town, from Los Angeles to Bangkok, and were it not for Cold War politics, from

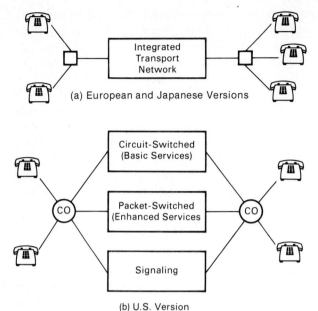

FIGURE 8-8
Two Versions of an ISDN: United States and Europe.

Chicago to Moscow. There are no problems with matching signals from one country to another and even the bills seem to come out right, although calls placed to some countries are more expensive than those placed to others. This ease of calling is the result of standards set by CCITT (the International Telegraph and Telephone Consultative Committee) that have been adopted by every nation wishing to communicate with another.

When complex communication systems have common or similar architectures, it is relatively easy to establish standards that allow systems to talk to one another. But now that there are significant differences between the U.S. telephone industry structure and that which appears to be emerging throughout the world, it is feared that the ability to communicate across nations will become more difficult. Certainly, creative standards will be required to ensure that the world is not disconnected. This is the task of the CCITT.

Of greater concern to the U.S. and to other nations with high technology industries is that telecommunications hardware and systems developed for the U.S. market will be unable to operate on foreign networks and vice versa, thereby creating serious trade barriers, barriers in that very industry that is considered to be one of the world's fastest growing.

The international politics of the telephone are the politics of trade. For if the PT&T's of the world provide their own forms of ISDN, there will continue to be little opportunity for trade in telephones and terminal equipment. However, many nations, and in particular the United States, seek freer trade and the opportunity to make terminal equipment available to what is probably the most lucrative business of the future, telecommunications equipment and information services.

While there may be good technical and economic reasons for embarking on the all-digital path, many believe that the PT&Ts also have political motives for their drive to the ISDN. High-valued telecommunications services in many nations help defray the operating costs of postal services, provide jobs, and serve as a reminder to a nation of its sovereignty. For these reasons, domestic and international telephone charges tend to be higher overseas when compared to the rates imposed by U.S. carriers. There is good reason, then, why a nation might wish to maintain its PT&T as a monopoly. One way of doing so is to ensure that all telephone hardware, from office to office and from home to home, is provided by the telephone monopoly and its local suppliers. Providing an ISDN with all of the intelligence built into the network where there is no differentiation between basic telephone services and such enhanced services as those requiring transmission digital signals, preserves the carrier monopoly and helps keep out foreign competition.

During the next several years, the CCITT will meet to decide upon standards for the telephone. The committee's deliberations go far beyond making certain that equipment and protocols are compatible. The politics of world trade will be on the table in Geneva as well.

TAKING STOCK

Let us pause for a moment to examine just what has been taking place in the technology of the telephone. You must agree that *time-space switching* is an excellent example of how the multitalented semiconductor chip has revolutionized the very nature of telecommunications and the convergence of computing and communicating. Without the chip, high-speed sampling would not have been feasible. Without the microprocessors made possible by that chip, digital encoding of the signal would never have occurred. Finally, without the semiconductor memories required to store the samples in order for them to be time-delayed, space switching would not have been possible.

Without one another, computing and telecommunicating are of limited value to modern society; their interaction has altered the shape of the industrial world and probably our lives as well. The emergence of a trade environment based upon telecommunications and information systems and services is of great political importance; there is hardly a nation in the world that does not see at least a portion of its future prosperity bound up with telecommunications and information. Vast energy resources are not required to enter this market. What is required are education and training. It has been remarked, more than once, that the Japanese miracle was sparked by Japan's decision in the nineteenth century to invest heavily in education.

REFERENCES

1 Dordick, et al., *The Emerging Network Marketplace* (Norwood, NJ: Ablex Publishing Corporation, 1981), pp. 39–43.

2 de Sola Pool, Ithiel, Hiroshi Inose, Nozumu Takasaki, and Roger Hurwitz, *Communication Flows: A Census in the United States and Japan* (Tokyo: University of Tokyo Press, 1984), pp. 109–120.

ADDITIONAL READINGS

Bellamy, John C., *Digital Telephony* (New York: John Wiley & Sons, 1982). The standard work on digital telephony; technical but very much worth scanning.

Brown, Paul B., Gunter N. Franz, and Howard Moraff, *Electronics for the Modern Scientist* (New York: Elsevier, 1982), pp. 409–426 on modems and digital signal processing.

Crane, Rhonda J., *The Politics of International Standards: France and the Color TV War* (Norwood, NJ: Ablex Publishing Corporation, 1979). While primarily concerned with television, this little book is an excellent example of the international politics of standards.

Doll, Dixon R., *Data Communications: Facilities, Networks, and Systems* (New York: John Wiley & Sons, Inc., 1978), pp. 139–218 on "transmission" and pp. 345–426 on "protocols and controls."

Miller, Gary M., *Handbook of Electronic Communication* (Englewood Cliffs, NJ: Prentice-Hall, Inc., 1979), pp. 231–265, for a good introduction to digital communications.

Williams, Frederick, and Herbert S. Dordick, *The Executive's Guide to Information Technology: How to Increase Your Competitive Edge* (New York: John Wiley & Sons, Inc., 1983). Part II, pp. 23–157, on how executives can make effective use of the intelligent telephone is particularly useful.

PART FOUR

THE SECOND ERA OF BROADCASTING

The ancients were fascinated by the strange rocks they found in Magnesia, rocks that could exert a pull or push away similar rocks without touching. The mystery of touching without touching—action at a distance—has always fascinated man. Maybe that is one reason telekinesis is a standard prop man's task in almost every horror film coming out of Hollywood.

Perhaps we have achieved the ultimate in telekinesis with broadcasting; radio and television have for more than sixty years played a most significant role in shaping society. Orson Welles created hysteria among more than 1.2 million listeners in 1938 with his fantasy about a Martian invasion; hundreds took to the streets in fear. What more can we ask for action at a distance? Television has been called a form of large-scale communion or fellowship, a media event that is greater than the event itself.

In Part 4 we explore broadcast technology and the new versions of broadcasting that have emerged in recent years. So important have been the changes in this technology that we have coined a new word to define a new form of broadcasting-narrowcasting. Cable television is just one of the several narrowcast technologies that is changing the nature of broadcasting, so much so that we call this "the second era of broadcasting." Perhaps the ultimate in narrowcasting is the radio telephone which marries one-to-one communications to the electromagnetic spectrum, to action at a distance.

In Part 4 we explore broadcast and narrowcast technologies, radio, television, cable television, and how new places in the spectrum have been found for multipoint distribution, for subscription television, and for satellite master antenna television. We conclude with mobile radio, an outgrowth of the convergence of broadcasting and the telephone.

CHAPTER 9

BROADCAST AND CABLE SYSTEMS: THE SECOND ERA OF BROADCASTING

THE MANY IMAGES OF BROADCASTING

Broadcasting is many things to many people. To the viewer or listener it is an entertainment medium, a convenient, low cost, and low personal energy consuming way to while away time or keep one's mind off the terrors of the highway. To the marketer and salesperson it is a way to capture the eyes and ears of millions of people when they are most vulnerable to advertising. To the creative artist it is a way of getting thoughts and messages delivered to a vast audience, a means for exhibiting hard-earned talents and getting well paid for it. One evening of Romeo and Juliet on television reaches an audience far larger than Shakespeare reached in his lifetime. To public officials it is means for keeping in touch with the electorate, a way of informing a democratic society of issues with which they need to deal. To educators it could be, and indeed in some areas of the world is, a powerful assistant. To many station license holders broadcasting is a way of making a great deal of money, and to some an activity with marginal financial but great psychic rewards.

Broadcast systems are public systems with enormous power to influence, persuade, and inform. It is no surprise that broadcasting has occupied a pre-eminent place on the world's political stage. National planners, politicians, educators, and creative artists, as well as the people on the street and on the farms, both love and hate, fear and trust broadcasting. Broadcasting and the words "mass communications" imply a sense of the new; in fact, nations have established radio and television broadcast systems as both a means for and a sign of modernization.

Radio waves know no national boundaries and do not step aside for local culture; the messages transmitted reach everyone democratically. Images of

Hollywood reach into Arabian deserts and James Bond is well known in Thai villages. *Kojak* and *M*A*S*H* have seduced the capitals of the world, overcoming language barriers with images of cars racing down Broadway and wars fought in the hills on the outskirts of Los Angeles. It was reported that the Israeli Kenesett adjourned a meeting earlier than planned in order to allow its members to see the outcome of a two-part episode of *Kojak*. In Indonesia, farmers now sell a plot of land to buy a television set rather than finance a trip to Mecca.

When the social impacts of media are discussed, it is television about which we are usually talking. The social impacts of television have been debated voluminously: television and children, television and violence, television and politics—all have been and continue to be major themes of research and literature among politicians, social scientists, psychologists, advertising executives, and government bureaucrats. We could easily begin a heated but endless debate among those who argue that television is dangerous to our health and those who counter with the argument that television is a positive force in our lives. We shall not get into this debate, for it would require many volumes just to state the established positions on both sides. We choose to remain neutral; whatever arguments are made, they are "not proven."

Broadcasting is universal; there is no nation in the world, rich or poor, that is not now seeking to deliver some form of broadcasting—radio or television—to its people. It is often political suicide to deny radio or television to the populace. For most of us, television has played an integral part in our lives; we are the television generation, born and raised by the dim blue light of the television screen. Beginning in the 1950s, the electronic media of radio and television rapidly spread from the industrial countries of Europe and North America to the developing countries of the world. There is no nation on the globe that is not within the beam of a radio signal, eagerly tuned to regardless of the quality of the signal. There is hardly a generation of young people throughout the world who has not heard a Beatles tune. Indeed, a popular song is already being hummed in the jungles of Irian Jaya on Papua New Guinea three days after its first playing in Djakarta.

Broadcasting and especially television is in transition, some might even say turmoil. This transition has been generated by new technologies and the opportunities they present for better achieving established broadcast objectives or going after new ones. The very idea of broadcasting is undergoing change. With the advent of cable television, the distribution of multiple video signals through cables, the availability of broadcast quality distribution channels that heretofore were denied to broadcasters, and direct broadcasting via satellite, we must coin new words to describe this second era of broadcasting, words such as narrowcasting and subscription television or pay-TV.

The technology works the same everywhere, but what it does is vastly different. Just as we found that the telephone is a purposive technology—we can do what we want with that technology—so can we view broadcasting as purposive. The difference is that nations rather than individuals make the choices. Even in those few countries where broadcasting is privately owned and driven by profit, national policies have been established to make the "most appropriate" use of the *electromagnetic spectrum*. What these technologies can do is determined by what the nation wishes to achieve.

In this chapter we examine the broadcast technologies. We explore how words and pictures are transformed into electrical signals, transmitted by electromagnetic radiation, and re-created to entertain and inform people and governments. We examine how improved technology, including the multitalented semiconductor chip, has changed the nature of broadcasting by uncovering new space in that finite natural resource, the radio spectrum, and creating new ways for delivering words and pictures through the cable as well as over-the-air. Finally, we explore why the future of television could be radio and why it might not.

ALLOCATING AND MANAGING SCARCITY

In our chapters on wired communications we learned how to enhance *connectivity*, how to use technology to bridge time and space for large numbers of people and places. Broadcasting requires that we create *separations* in time and space so that different messages can be heard and seen by many groups of people without interference from one another.

Economic and effective utilization of bandwidth in our wired systems conserves copper, underground ducts, and unsightly overhead wires. We are in no imminent danger of exhausting nature's supply of copper, but in our search for lower cost transmission alternatives, glass in the form of optical fibers is now being substituted for copper. We are not likely to exhaust the supply of sand from which to fashion glass fibers. Indeed, there are no natural bandwidth limitations to wired telecommunications, only limitations in our ability to design equipment that can use ever and ever higher frequencies.

Broadcasting, on the other hand, utilizes a natural resource known as the *electromagnetic spectrum*, which we examine in this section. If it is a limited natural resource, how come we have "discovered" new spaces for new broadcast services?

The Electromagnetic Spectrum

In Chapter 2 we described how information can be sent without wires, or broadcast. We found that in order to do so we must vary an electric charge or the current in a wire to create a disturbance in the surrounding atmosphere (called an electromagnetic field). The disturbances created are electromagnetic waves which contain the information transmitted. These waves and the field in which they move extend to infinity. It is convenient to categorize the electromagnetic fields created in terms of frequency, since they are waves and we have seen that one way to measure a wave is in terms of its frequency of vibration or number of cycles per second.

The tremendous range of frequencies produced by an electromagnetic disturbance is called the electromagnetic spectrum.

Perhaps the high point in the physics of electricity did not occur in the laboratory but on a pad of paper on the desk of James Clerk Maxwell, the brilliant British physicist/mathematician we encountered in Chapter 2. In 1873 he argued that an oscillating electrical circuit (our current in the wire, for example) should radiate electromagnetic waves. Twenty years later Heinrich Hertz proved that this was indeed so using a device that generated a rapidly varying or oscillating current; today we call that device an oscillator. Clerk Maxwell calculated that the velocity of the waves was very nearly that of the velocity of light. The evidence seemed inescapable that light was also an electromagnetic wave. But how did the light waves differ from electromagnetic waves?

This brings us to another way of categorizing waves—that is, by their wavelength. Let us refer back to Figure 2-5(a), to the sine wave which has become our major transportation vehicle for broadcasting information. Wavelength is defined, quite simply, as the distance between the successive peaks of the sine wave, points 1 and 2 on Figure 9-1.

Imagine now, that we attach a string or cord to the wall as in Figure 9-2 and produce waves in the string by moving one end up and down regularly once a second. The first wave we form moves down the string toward the wall, the next one behind it, and so on. The waves travel at a constant velocity (if we wave regularly) and as they are formed, they move down the string a distance of one wavelength in one second. To put it another way, the velocity of the wave is equal to the distance traveled per unit of time. But since time is inversely related to frequency (remember, cycles per second) the speed of a wave (or the velocity of propagation, to put it precisely) is equal to the frequency times the wavelength.

Knowing that the speed of light is constant, Clerk Maxwell reasoned that light waves are electromagnetic waves with very short wavelengths. In his laboratory twenty years later, Hertz proved Clerk Maxwell right.

Thus the electromagnetic spectrum can be defined both in terms of frequency and wavelength. The electromagnetic spectrum runs from zero frequency or direct current with an infinite wavelength to the frequency of your electrical appliances at 60 cycles per second with a wavelength of just about 5 million meters long; to your AM radio with a frequency of about a megahertz

FIGURE 9-1
Wavelength of the Sine Wave.

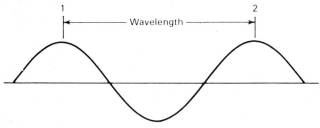

and a wavelength of 300 meters; to shortwave radio with a frequency of 10 megahertz and a wavelength of 30 meters; on through television, infrared, visible light with a frequency of a thousand million million cycles per second and a wavelength of a hundred millionth of a meter; and so on to the cosmic photons from astronomical sources with frequencies in the range of 10^{23} (1 followed by twenty-three zeros) and wavelengths of 10^{-15} (1 preceded by fifteen zeroes) as shown in Table 9-1. (You can calculate your own wavelengths for any frequency you select simply by remembering that the velocity of light and all electromagnetic waves is a constant 186,000 miles per second, or about 300,000 kilometers per second, and that the velocity of an electromagnetic wave is equal to its frequency multiplied by its wavelength.)

The Radio Spectrum

The electromagnetic spectrum is one of our irreplaceable natural resources; once filled with uses, it cannot be expanded. We have touched on ways to better utilize this spectrum in our discussions of efficient use of bandwidth through sharpening the boundaries between frequency multiplexed channels and the use of time division multiplexing. The reason we need to use these bandwidth conservation techniques is that the portion of the spectrum we can use for broadcasting is, when compared to the entire spectrum, quite small, as shown in Table 9-1 and somewhat expanded in Table 9-2. It ranges from very low frequencies of a few kilohertz and wavelengths of several kilometers up to 300 gigahertz (billions of cycles per second) at which point radio microwaves move into the far infrared. (In this context, we use the term radio to encompass all forms of wireless communication, including television, radio broadcasting, and telephone calls sent by microwave radio and satellite.)

Radio waves can travel on or near the ground (the ground and direct waves), or they can travel upward some hundreds of kilometers where they are bent or reflected by a band of ionized particles trapped in space—the *ionosphere*—and bounced back to earth some distance away. These are the skywaves in Figure 9-3. The direct waves may travel in straight lines or *line-of-sight* between antennae over short distances, usually not more than 100 kilometers, or as ground waves reflected off the ground in order to reach the receiving antennae.

FIGURE 9-2
Velocity of a Wave.

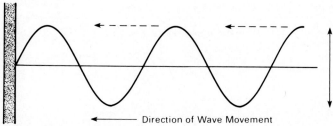

TABLE 9-1
THE ELECTROMAGNETIC SPECTRUM

Frequency in Hertz	Wavelength in Meters	Nomenclature	Typical source
10^{23}	3×10^{-15}	Cosmic photons	Astronomical
10^{22}	3×10^{-14}	γ-rays	Radioactive nuclei
10^{21}	3×10^{-13}	γ-rays, x-rays	
10^{20}	3×10^{-12}	x-rays	Atomic inner shell
		Positron-electron annihilation	
10^{19}	3×10^{-11}	Soft x-rays	Electron impact on a solid
10^{18}	3×10^{-10}	Ultraviolet, x-rays	Atoms in sparks
10^{17}	3×10^{-9}	Ultraviolet	Atoms in sparks and arcs
10^{16}	3×10^{-8}	Ultraviolet	Atoms in sparks and arcs
10^{15}	3×10^{-7}	Visible spectrum	Atoms, hot bodies, molecules
10^{14}	3×10^{-6}	Infrared	Hot bodies, molecules
10^{13}	3×10^{-5}	Infrared	Hot bodies, molecules
10^{12}	3×10^{-4}	Far-infrared	Hot bodies, molecules
10^{11}	3×10^{-3}	Microwaves	Electronic devices
10^{10}	3×10^{-2}	Microwaves, radar	Electronic devices
10^{9}	3×10^{-1}	Radar	Electronic devices
		Interstellar hydrogen	
10^{8}	3	Television, FM radio	Electronic devices
10^{7}	30	Short-wave radio	Electronic devices
10^{6}	300	AM radio	Electronic devices
10^{5}	3000	Long-wave radio	Electronic devices
10^{4}	3×10^{4}	Induction heating	Electronic devices
10^{3}	3×10^{5}		Electronic devices
100	3×10^{6}	Power	Rotating machinery
10	3×10^{7}	Power	Rotating machinery
1	3×10^{8}		Commutated direct current
0	Infinity	Direct current	Batteries

The Radio Spectrum: 10^{3} to 10^{12} Hz

Used with permission from Encyclopedia of Physics, p. 278, McGraw-Hill 1983.

TABLE 9-2
THE RADIO SPECTRUM

Frequency in Hertz	Wavelength in Meters	Designation
10^1	10^7	Extremely Low Frequency (ELF)
10^2	10^6	Voice Frequency (VF)
10^3	10^5	Very Low Frequency (VLF)
10^4	10^4	Low Frequency (LF)
10^5	10^3	Medium Frequency (MF)
10^6	10^2	High Frequency (HF)
10^7	10^1	Very High Frequency (VHF)
10^8	$10^0 (=1)$	Ultra High Frequency (UHF)
10^9	10^{-1}	Super High Frequency (SHF)
10^{10}	10^{-2}	Extremely High Frequency (EHF)
10^{11}	10^{-3}	Band No. 12
10^{12}	10^{-4}	

(Purists will add that radio waves also can travel through the earth by way of the rock strata.) Higher frequency signals, those in the very high frequency range (VHF) and ultra high frequency range (UHF) travel between antennae directly, or by line-of-sight.

How this radio spectrum is utilized is governed by the physics of radio wave propagation as just described and by the practical limits of communications engineering design. Let us consider just what is taking place when you sit on the beach with your portable radio listening to the ballgame. The sportscaster in the booth overlooking the field is shouting the play-by-play into the microphone which transforms his voice into electrical signals, much as the microphone in the telephone handset does. These signals are carried by cable from

FIGURE 9-3
How Radio Waves Reach You.

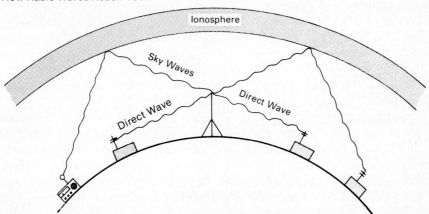

the booth to a transmitter which may be as much as twenty miles away from you. There, the signals modulate an electromagnetic wave created by a rapidly vibrating sinusoidal wave, the carrier at, say, 1070 kilohertz. The now-modulated carrier signal carrying the play-by-play is fed into the antenna from which it spreads out in all directions as electromagnetic radiation.

On the beach, twenty miles away from the transmitter, your transistor radio antenna is bombarded by signals from all of the radio stations broadcasting from that transmitting antenna as well as others perhaps on nearby mountaintops. Your radio selects a particular signal from the many based on the fact that you have tuned it to 1070KHz (station KNX in Los Angeles, for example).

The transmitter is sending out the signals in all directions at a power of 50,000 watts and if this power were spread out evenly, your transistor radio antenna would receive about 0.000003 watts of power. Actually, most of the power from the transmitted antenna is directed to the horizon so your transistor radio antenna might well be receiving only a millionth of a watt! This is the signal the amplifiers and demodulators in your radio have to work with and which they must amplify and demodulate. Most radio stations transmit at power levels far below that of our sample station, a CBS affiliate in the nation's largest radio market. Amplifiers and demodulators must work with very minute signals, indeed.

Life is not that simple, however, for in addition to sand in your face, you may not be hearing the plays very well; various kinds of interfering radio waves and other sources of noise make it difficult for your radio to pick up the signal you want. You will remember from Chapter 2 that many sources of noise surround you, including natural radiation-producing phenomena such as distant lightning and other happenings in the outer galaxies, manmade electromagnetic noise from badly tuned automobile engines and power tools, and from other radio transmitters that could be quite far away but still bombarding you with skywaves, and reflections from tall buildings. How often have you found a distant signal from as far away as a thousand miles interfering with your local station at night? That's because certain layers in the ionosphere provide the means for long-distance radio transmission, the means by which transAtlantic radio telegraph communications were carried out before the satellite.

Allocating the spectrum in order to minimize this interference and to create the separations we mentioned at the beginning of this chapter is the purpose of spectrum management. Regions of the spectrum have been set aside for specific purposes in order to allow for their most efficient use without interference. Thus, the band from 535KHz to 1605KHz is reserved for AM radio broadcasting.

In the early years of radio in the United States and in other nations, frequencies within this band were not assigned by any government agency. Indeed, in the United States several legal obstacles had to be overcome since the courts initially found that the Department of Commerce could not tell entrepreneurs where they could carry on their broadcasting business. Minimizing interference was left to the broadcasters themselves who adjusted their frequencies in order to avoid as much interference as possible. But this interference also depends on the power transmitted and on the natural geographical obstacles a radio signal might encounter. Mutual adjustment by the broadcasters did not work too well, especially when more money could be made by transmitting at one frequency than at another—for example, at a frequency at the center of the band where tuning is almost mechanical rather than at the extreme left or right which the listener must make some conscious effort to reach.

Assigning the radio spectrum to specific uses is a worldwide activity; the radio spectrum is a natural resource that must be shared by all nations and managed on a worldwide basis. Recall from our discussions of the satellite how it is possible to use the same frequency at the same time in different parts of the world. The same is true for other radio frequencies. Television Channel 4 is used in New York, Washington, DC, and about fifty other locations throughout the United States. The AM radio band, too, is shared by all of the world's broadcasters. The international institution that allocates places on the spectrum and promulgates technical standards and rules for radio spectrum use is the International Telecommunications Union (ITU), an agency of the United Nations. In the United States the responsibility for managing the radio spectrum is divided between the President in whose name spectrum assignments to agencies of the federal government are managed by the Department of Commerce and the Federal Communications Commission (FCC). The responsibility of the FCC includes the assignment of all radio activities from those for commercial broadcast and communications to those that operate the opening and closing of garage doors.

Not only can there be interference between radio spectrum users within the portion of the spectrum assigned to them, there also can be interference between different uses. For example, mobile radio interferes with television reception, but television does not interfere with mobile radio. There are more incentives for cooperation and coordination if a single class of users is assigned to each region of the spectrum. Consequently, the radio spectrum has been assigned for specific uses and to similar groups of users, as shown in Table 9-3.

Within these regions, the spectrum is managed by national organizations such as the FCC. The process is threefold: allocation, assignment, and licensing. The results of the allocation step are shown in Table 9-3. How the band is assigned varies. For example, the AM broadcast band is allocated on a first-come, first-served basis. A prospective user searches for an unused slot in the band and, finding one, registers an intent to use it. If the proposed use for that slot does not interfere with other users, the searcher will be licensed to use that frequency. On the other hand, the FCC worked out a table of assignments for television before opening the band to licensees in order to simplify and speed up the licensing process and to ensure that its goal of localism would be met by balancing the assignments to communities.

You will recall the U.S. policy on orbital slot allocations for satellites. Our nation opposes the Third World model for a priori assignments, which is exactly what we do for television and for various other domestic frequency assignments. Consistency in policy making is not one of this nation's strong points.

TABLE 9-3
THE USES OF THE RADIO SPECTRUM

Frequency	Designation	Radio Use
10^4	Low Frequency	Long wave radio formerly for transatlantic telephone
10^5		
	Medium Frequency	AM Broadcasting
10^6		
	High Frequency	Short Wave Radio Ship-to-shore radio Telephone
10^7		
	Very High Frequency	FM Broadcasting VHF Television
10^8		
	Ultra High Frequency	UHF Television
10^9		
	Super High Frequency	Microwave Relays Satellite Communications
10^{10}		
	Extremely High Frequency	
10^{11}		

Even if the number of available channels far outstripped the demand for their use, we would still have to coordinate the allocation of the channels to avoid interference. However, whenever a government has the opportunity to allocate scarce resources, it can control how these resources are being used. Thus spectrum management turns out to be a way of making broadcast policy.

The FCC sought to ensure balance in the distribution of the limited number of television channels in the assigned band of the radio spectrum, so that no region of the nation would be left out of the marketplace of televised ideas. The spectrum management process also must allocate scarcity and thus set certain requirements for broadcasting that fly in the face of our usual goals for entrepreneurial freedom and the First Amendment. There are limits placed on the number of broadcast stations any single firm can own and broadcasters are required to provide balanced reporting of different points of view, a rule that does not apply to newspaper, magazine, or book publishers. The U.S. Supreme Court has supported this policy on the grounds that there is scarcity in broadcast frequencies and that because access to information is important to a democracy, the government should have the power to set rules for the use of the radio spectrum. Clearly, in nations in which broadcasting is government owned and operated managing the spectrum can be a way to control who says what to whom.

The rules are changing all over the world; the satellite, the computer, and the semiconductor chip have made possible more efficient uses of the radio spectrum, much to the chagrin of those who came first and are now in business. In the United States, additional broadcast entrants could cut into the audience numbers a competing program captures, thereby reducing the value of that program to the advertiser. Remember, it is the advertiser who pays the bills. In other nations, additional channels create opportunities for new revenues to support national programming but also generate pressures from the public for more diverse programming, from the advertisers for more and better programs on which to advertise their products, and from a new breed of broadcast entrepreneurs to allow wider access to the broadcast business.

Now we examine just how and why the technology is making all of this happen.

TRADITIONAL BROADCAST TECHNOLOGY

In our earlier discussion of wireless communications we noted that in order to send signals over long distances by means of wireless transmission or broadcasting a carrier is needed on which to superimpose our information signal. This carrier has to be at a much higher frequency than that of the human voice. To generate the carrier, a radio frequency oscillator, a device that can create rapidly moving sinusoidal carrier waves, is required.

We also discussed the processes of modulation and demodulation. Now we put these functions together to "design" a radio broadcast system.

Radio Broadcasting

The oscillator in Figure 9-4 generates a radio frequency (RF) carrier which is modulated by the signal carrying the information. This process is called *mixing*. The combined signal, or modulated carrier, is very low in power, perhaps on

FIGURE 9-4
The Basic Broadcast System.

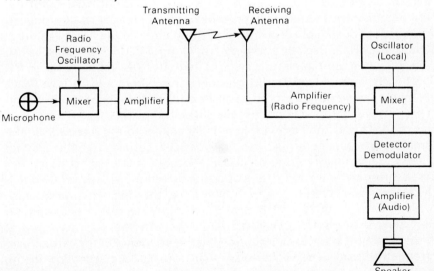

the order of milliwatts or thousandths of watts. The amplifier increases the power to the 50,000 watts required to deliver the ballgame to you at the beach.

At the receiving end, the sequence of events is reversed. The receiving antenna picks up the now very weak signal and amplifies it so that it can be processed for detection. The detector is the demodulator which extracts the intelligence from the modulated carrier. We now have an audio signal—the play-by-play of the ballgame—and it must be amplified so that it can operate or "drive" the speaker in our radio.

The first radio receivers for AM broadcasting, called tuned radio frequency or TRF receivers, were very simple. Two important characteristics of these or any receivers are their sensitivity and selectivity. Noise is always present in any transmission of signals, whether over-the-air or through a wire; when we amplify a signal, we also amplify the noise. (The exception, as noted in Chapter 2, occurs when we send digital signals; in such cases, we do not amplify, we regenerate the pulses.) Well-designed radio receivers are capable of amplifying the smallest of signals, from thousandths of a volt in low-cost radios to millionths of a volt in the higher-priced radios with good signal to noise ratios. It is safe to say that one of the most important qualities you pay for in any radio frequency receiving device, including television receivers, is the ability to pick up and amplify, with good quality, extremely weak signals.

Another important quality about a receiver is its selectivity, its ability to differentiate the desired signal from the other signals and disturbances in the "air." The better the selectivity, the higher the cost. Radio pioneers would spend hours carefully tuning their crystals and their many amplifiers in order to detect the signals they wanted.

The quality of amplification depends on the range of frequencies of the signals being amplified. Your high-fidelity music systems are designed to amplify signals over a wide range of frequencies (usually between 20 cycles per second to about 20,000 cycles per second), but to do so these systems are quite complex and often extremely expensive. Radios could not have swept the world if amplifiers had to operate across wide frequency ranges. Mass-produced radios must be capable of receiving all of the stations in the AM band, from 535KHz to 1605KHz and sell for anywhere from $10 to $150 for the multispeaker, cassette-included, portable music systems that often disrupt our visits to the park or beach. Preferably, the radio frequency amplifier should do its work at a single frequency; its design should be optimized for that single frequency and not require several tuning knobs to tune several amplifiers.

This problem was solved by using a mixer similar to the one used in the transmitter, to beat or *heterodyne* the carrier signal against a signal at a higher frequency created by the local oscillator. Heterodyning a carrier frequency against a sinusoidal wave running in opposite directions (simply, when one is up the other is down) results in a frequency lower than the carrier frequency. The resulting frequency, called *intermediate frequency,* remains constant no matter what the carrier frequency happens to be. (Another way of looking at "beating" is simply as a means of subtracting the frequency of the carrier from the frequency produced by the local oscillator.) For the reception in the commercial AM radio band to occur, the local oscillator is always tuned to 455KHz above the frequency of the carrier as broadcast by the station to which you are tuned. At the conclusion of the mix, the intermediate frequency amplifier will always operate with a center frequency of 455KHz. This is the famous *superheterodyne receiver* design principle invented in the 1930s and still found in most receivers today (see Figure 9-5).

Finding New Broadcast Space

Each AM radio channel (and broadcaster) is allocated a 10KHz band. For example, our ballgame is being transmitted on Los Angeles station KNX which is assigned a center carrier frequency of 1070KHz in order to operate in the assigned range of 1065KHz to 1075KHz. The play-by-play occupies a maximum of 5KHz (B in Figure 2-13(a)), which generates an upper and a lower sideband, F_c+B and F_c-B respectively in Figure 2-13(c) where F_c is the center carrier frequency 1070KHz.

An amplitude modulated carrier generates a signal consisting of three different frequencies: 1) the original carrier with the amplitude unchanged; 2) a frequency equal to the difference between the carrier and the modulating frequency; and 3) a frequency equal to the sum of the carrier and the modulating frequency. When the incoming radio signal is demodulated, we remove the carrier and capture the information on the sidebands. In the early 1920s experiments showed that we could remove one of the sidebands entirely and still capture the information on the other sideband. Furthermore, we found it possible to transmit both sidebands with different information on each, demodulate or supress the carrier and broadcast two messages on a single

FIGURE 9-5
The Superheterodyne Receiver in Principle: Converting Frequencies.

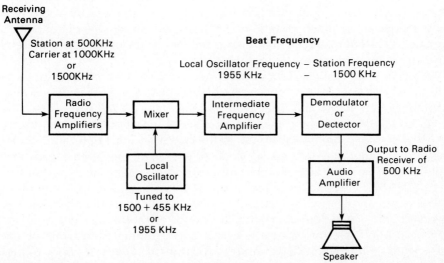

radio station. By 1923 the first patent for a *single sideband* (SSB) radio system was granted.

Here is the first technological development that opened up "new" bandwidth. Because SSB transmission conserves bandwidth by making two signals available on one assigned carrier, the FCC is requiring its use in the overcrowded 2–30MHz range.

In the AM band between 535KHz and 1605KHz there can be 107 AM broadcast channels each with a usable bandwidth of 10KHz. High-fidelity sound cannot be expected from a 10KHz channel, indeed, many other countries have only 8KHz. In any case, FM is less susceptible to noise and is more suitable for high-quality or high-fidelity sound broadcasting. The FCC recognized this potential for FM transmission and assigned each FM station a bandwidth twenty times greater than assigned to AM radio—200KHz in the station's own band between 88MHz to 108MHz. Today, most radio receivers are both AM and FM. With FM available for high-quality sound reception, the FCC proposed to reduce the assigned bandwidth for AM radio to 9KHz thereby making room for as many as ten additional stations in a market. The FCC also argued that improved design of electronic equipment could result in much better utilization of the narrower bandwidth.

You ought not be surprised to learn that the politically powerful broadcast industry did not receive this proposal with much enthusiasm; new entrants would mean more competition in a market they believed was already too crowded. So we see that technology does not always carry the day, even if the objectives are noble.

In the 1920s when radio first burst on the scene, radio frequency oscillators could generate only low frequencies, usually not more than 300KHz. Nor were there devices capable of modulating and amplifying high frequency carriers and signals. The history of broadcasting is the story of the constant and continuing search for hardware improvements that enable the use of higher speeds and bandwidth. In the 120 or so years since Clerk Maxwell speculated that a rapidly changing electrical current in a wire—an oscillating current—would generate an electromagnetic spectrum, we are capable of generating signals with frequencies of 100 million million cycles per second, thereby opening up new bands for radio transmission.

When new technological tricks are developed and uses are found for these new bands, prospective entrepreneurs will ask the FCC to assign bands or find room for them in the radio spectrum. The Commission will review how the assigned bands are being used. If the FCC finds one not being adequately utilized, or if the Commission comes to the conclusion that other uses would be more in the public interest, a "docket" will be opened. This means that opinions will be solicited from those now using the band and those wishing to use portions of the band for other purposes. With the rapid improvements in radio broadcast technology FCC intervention has happened with increasing regularity in recent years. In particular, rapid technological developments have had significant impacts on television in the United States and we are witnessing a veritable explosion of new over-the-air broadcast opportunities to challenge the long-held stability and profits of the broadcast networks. Similar trends are taking place in Europe and Japan where government-owned television networks now find themselves challenged.

We turn our attention now to television and see how technological advances are leading us to its second era. We begin with over-the-air, or broadcast, television, then examine cable television, or television by wire.

Television Broadcasting

There may be some among you who still remember the small prizes found in your Crackerjack box. One of these might have been a matchbook size bundle of pages with stick figures drawn on them. When you flipped through the pages these figures seemed to move. The faster you flipped the better the motion, the slower you flipped the more flickering the pictures became. If you have never been fortunate enough to have received this wonderful prize, take a small blank notebook and on the same spot on each page draw a stick figure whose posture changes slightly from page to page. Now riffle through this notebook and you, too, will have created a "motion picture."

The phenomenon that creates this apparent motion is "persistence of vision." The eye perceives each drawing and retains the individual impression for a period of time that is long enough to overlap to the next picture. The retina of the eye can hold each image for about 1/30 of a second. If we flip the pages of our wondrous matchbook at a faster speed, the pictures will be seen not individually but as a blend the eye perceives as motion.

The persistence of vision is what makes motion picture cartoons possible; a ten-minute Mickey Mouse cartoon is composed of 14,400 individual drawings, each slightly different from the one preceding it in the sequence in which they are presented. The finished product you view on the screen is the continuous flow of these drawings at a rate of twenty-four pictures or frames per second or 1440 frames per minute. Since the eye persists in holding on to the image for only 1/30 of a second, projectors use a shutter to cut off the stream of light from the image twice during the projection of each frame so that forty-eight separate pictures are really projected each second.

It is likely that this characteristic of our eyes created the urge to somehow "store" and recapture pictures in motion just as we capture still pictures with photographs. The discovery of the light sensitivity of selenium, a semiconducting device that preceded the wonderful chip (see Chapter 3), and the invention of the telephone in the late 1800s set the stage for serious consideration of the possibility of recording moving visual images electronically and sending them over distances first by wire and, after Clerk Maxwell, by radio. In 1884, Paul Nipkow patented a spinning disk scanning system for television. This device had a spiral array of holes. A selenium photocell behind the disk would convert the line-by-line tracings of light patterns into a signal which could be transmitted by wire to a display device. There it would modulate the intensity of a light source behind another spinning disk with the same pattern of holes. The viewer would see the original picture reproduced element by element.

The disk would spin fast enough to satisfy the eye's persistent desire to "remember" the previous image and fuse it with the next one.

By the 1920s, the British Broadcasting Company (BBC) was using this mechanical system to broadcast television after normal broadcast hours. The Bell Laboratories improved on this system, providing regular television broadcasting as early as 1927 and the viewing audience began to grow. But these mechanical systems were large, expensive, and difficult to maintain. The key to the success of television was the switch from mechanical scanning to electronic scanning using the cathode ray tube invented by Braun in 1897 and the iconoscope camera tube patented by Zworykin in 1923.

It is interesting to note that color television followed the same pattern of development. Mechanical systems were introduced in the 1920s culminating with the Goldmark-CBS system which used a disk containing color filters spinning in front of a conventional black and white receiver. The system worked fine as long as you could synchronize the receiving disk with the colors called for by the transmitter. The virtue of the Goldmark-CBS system was its compatibility with the many black and white television sets on the market.

An all-electronic system was under development by RCA at the same time and while this later replaced the bulky mechanical system, it required special receivers, certainly a plus for the company that manufactured television receivers. The RCA system at first used three separate camera tubes and three separate picture tubes each with its own color filter and mirrors to mix the three primary colors. Today the three tubes have been combined into one with accompanying devices for mixing the three primary colors in accordance with the information being transmitted. We shall see how this is accomplished in the sections that follow.

In the previous section we described a radio frequency broadcast system and Figure 9-4 illustrates just such a system. The transmission system for television is generally the same except that the input and output signals are singularly different. The radio technology described previously transmits either voice, music, or data. The input signals for television are audio *and* video and there must be appropriate devices in the system for matching up the complex signals that are sent and received. We have had some experience with the matching or synchronization process for the transmission of data in the previous chapter. Making certain that video signals representing many varying shades of black and white or hues of color are received as transmitted requires devices that respond to more complex instructions than do those that recognize the existence or absence of a pulse. Add to this the transmission of voice synchronized with the picture and you have a most complex electronic system. Indeed, the television set in your home is one of the marvels of engineering; it is certainly one of our most complex devices yet it is easily controlled and operated by children of all ages and is remarkably trouble free.

The television camera is optically focused so that the scene to be transmitted appears on its light-sensitive area, a photosensitive semiconductive material in the form of a matrix or mosaic as illustrated in Figure 9-6. Let us suppose that we focus the camera on a letter "A" in Figure 9-6(a) and that we scan this letter

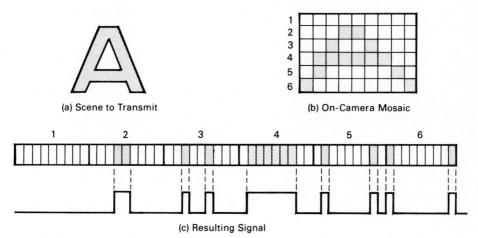

FIGURE 9-6
How a Scene is Scanned Electronically.

with our camera. The letter is brightly lit on a dark background. The semiconductive material is light sensitive—that is, it produces a signal proportionate to the intensity of the light. The mosaic will be illuminated as shown in Figure 9-6(b); the letter "A" is lit while the remainder of the mosaic is dark. Since light on the mosaic will generate a current and dark will not, the resulting video signal will appear as in Figure 9-6(c).

Now let us see what we have to do to generate a video signal that is much richer than our letter "A" and is moving. The camera converts the large and moving visual image into the required electronic signal by repeatedly scanning across the image from left to right in a series of horizontal sweeps starting at the top and ending at the bottom. During each sweep, the instantaneous light intensity is converted into an electrical current as with our letter "A" above. The number of sweeps necessary to reproduce the picture is one of the tradeoff decisions with which communications engineers must so often deal.

The more sweeps the better the picture, but the more sweeps the more rapidly the transmitted signal must move. We found in Chapter 2 that the more rapidly a signal changes, the more bandwidth is required to transmit that signal. To ensure every community in the country access to local television we must conserve bandwidth. In the United States a compromise has been reached that satisfies the viewers and takes up the smallest possible bandwidth to achieve viewer satisfaction. This has turned out to be the 525-line scanning pattern shown in Figure 9-7.

The 525 lines are scanned in two waves—lines 1 to 262.5 in one set of sweeps and lines 263 to 525 in the second set of sweeps. Together, the two interlaced sweeps produce a frame; thirty frames are produced each second.

This complex set of activities must be faithfully reproduced in the receiver; consequently a number of control pulses must be created that will synchronize

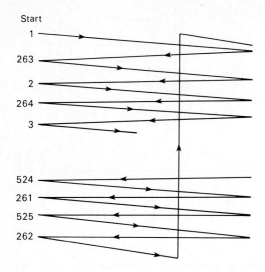

FIGURE 9-7
The 525-Line Scanning Pattern.

the transmission activities at the receiver. A simplified version of the complete video signal or the composite video signal is shown in Figure 9-8. Triggering the scan lines from left to right and sending them back from right to left is the job of the horizontal synch pulse. Sending the scan from the left bottom of the screen to the top right in order to begin the horizontal scanning is the job of the vertical synch pulse. The equalizing pulses keep the scanning mechanism under control.

One very important fact about the composite video signal should be noted: not all of the time throughout the signal transmission is taken up with video intelligence. There is a period when the scanner goes back up to start all over again; this is the vertical blanking interval when neither pulses nor information are being transmitted. During this period much of the assigned bandwidth is unused. This bandwidth is valuable and in a well-managed spectrum, it should be utilized. In Chapter 10 on publishing without paper we will see how no

FIGURE 9-8
A Simplified Composite Video Signal.

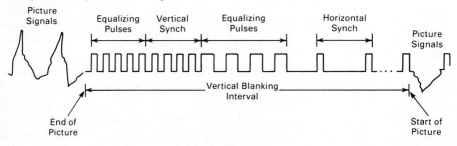

unfilled bandwidth goes unused and have more to say about the valuable vertical blanking interval.

The bandwidth required to transmit the composite video signal—a color television signal—is 6MHz. The FCC has allocated several bands for television:

54MHz to 75MHz for television channels 2, 3, and 4;
76MHz to 88MHz for television channels 5 and 6;
174MHz to 216MHz for television channels 7 to 13; and
470MHz to 806MHz for the UHF television channels 14 to 69.

However, these are not the only television bands set aside by the FCC. In the 1960s during the optimistic heydays of educational television, the band between 2500MHz and 2690MHz was set aside for use by schools and colleges. Because of the frequencies of signals in this band and, of course, their very small wavelengths, Instructional Television Field Services (ITFS) signals had ranges limited by their line-of-sight, usually a range of twenty-five miles, and were not available to the regular television receiver. Reception of ITFS broadcasts required a *downconverter* that would convert the high frequency signals to the lower frequencies that normal television receivers can use. Unfortunately, downconverters are not within the price range of most home viewers, certainly not those who might wish to receive instructional broadcasts. We shall have occasion to return to a discussion of the ITFS band a bit later in this chapter.

So far we have described only the video signal; the television signal is composed of both voice and video. The maximum modulating signal for the video signal is actually 4MHz. Since it is amplitude modulated on the carrier, this should result in a bandwidth of 8MHz, 4MHz on each side of the carrier center frequency. However, since the FCC allows only 6MHz for the television signal and that includes the FM audio signal, some adjustments must be made, as shown in Figure 9-9. Only 1.25MHz of video signal is carried below the

FIGURE 9-9
The Transmitted Television Signal: Sound and Picture.

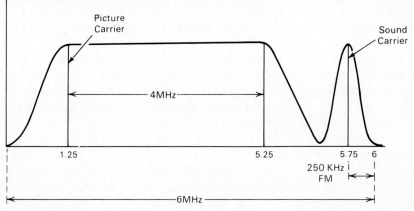

carrier, the rest is filtered out. The upper sideband of 4MHz is carried in full. The entire FM audio, upper and lower sidebands, is carried. We shall have occasion to refer to these sidebands shortly because they, too, represent portions of the spectrum that might be better utilized.

BREAKING UP THE OLD CLUB: THE NEW TELEVISION

During the past fifteen or so years, the well-established and politically powerful U.S. broadcast industry has been shocked out of the profitable state of lethargy it has enjoyed for more than forty years. The industry's profits have grown enormously, and if you believe the "ratings," so has its popularity. The polls always show broadcast anchormen and anchorwomen to be among the most credible of news and opinion sources; indeed, they are considered the spokespersons for our nation. The popularity of television personalities rivals that of early movie stars. The average household reportedly views television six hours and forty-eight minutes per day and there seems to be no limit to the public's appetite for video.

Radio which was supposed to have died with the birth of television continues to thrive; no jogger worth his or her running shoes could be without a "Walkman" and the commuter's sanity in traffic jams can only be salvaged by listening to the radio.

However, as television's power and popularity grew so did questions about its impact on society. Foundation and academic research reports vied with those coming from advertising agencies and the networks themselves in condemning or defending television. The federal government entered the fray with the *Surgeon General's Report on Television and Violence*—not the first time the government had examined a mass medium. In 1938 the Payne Committee researched the impact of movies on children and reached conclusions that were, in many ways, similar to those reached by the Surgeon General. Television, the Surgeon General argued, has contributed and continues to contribute many positive features to society; it stimulates learning through creative and often very original children's programming, and it stimulates the desire to read through programs that re-create the past and dramatize great literature. Television can expand the visions of young and old alike by bringing art, literature, music, opera, and drama to larger audiences than any medium in history.

But television also can create visions that are far from reality, skew the viewer's images of people and places in a way that is entirely unreal. Almost since its introduction in the 1950s, many have feared television's impact on the very young. While it cannot be *conclusively* shown that television has negative impacts on children, few would contest that such heavy television viewing must have some impact and it may not be entirely beneficial; indeed, it is conceivable that television could be quite damaging to a child's mental and physical health. While we shall not enter into this argument, it is important to note that the public's doubts about television seemed to emerge just when improved broadcast technologies began to appear offering the prospect of a significant alteration in the shape of the nation's television institutions.

In this section we examine these new technologies and find out just how they are altering the nation's broadcast structure. In so doing, we shall discover that the boundary between the telephone and the television set appears to be breaking down and perhaps we may need to redefine just what is meant by the term "mass media."

Cable Television

In our discussion of how the plain old telephone can be used to transmit data (Chapter 8) we noted that the twisted pair copper wire could deliver frequencies up to and perhaps a bit above 54,000 hertz (54KHz). An AM radio channel has a bandwidth of 10KHz with only half being used to transmit voice or music. With frequency multiplexing the copper wire could transmit as many as ten AM broadcast channels. If all of the wire could be used for the transmission of music, the fidelity of wired radio could be viewed as being about 1/4 that of the FM radio transmission. Wired radio, which today would be called cable radio, was quite popular in many European countries. To this day, in fact, many hotels in Iron Curtain countries continue to provide local radio by wire.

The copper wire of the telephone twisted pair cannot, however, deliver signals which require larger bandwidths, such as the 6MHz required for color television. The reason for this is the atomic nature of electrons and protons and what the physicists call the *skin effect*.

When vibrating electrons travel along the copper wire, they tend to congregate on the surface of the wire since they repel each other because of their same polarity, as illustrated in Figure 9-10(a). As the vibrations increase—that is, as the frequency of the current increases—the surface of the wire becomes very crowded and the electrons have difficulty moving; consequently, the *conductivity* of the wire decreases. If additional surfaces could be provided on which the electrons could congregate, we could increase the frequency while not decreasing the conductivity.

Enter the *coaxial cable* shown in Figure 9-10(b). A coaxial cable consists of two conductors, an inner conductor of copper surrounded by a foam insulator and an outer conductor of either copper or aluminum covered by a plastic jacket. The coaxial cable provides three surfaces on which the electrons can congregate and travel. With three surfaces, the cable can accomodate much higher bandwidths than on the one surface copper wire. It was the marriage of this coaxial cable with television that created the cable television industry, along with the imagination of an appliance store owner in the hills of Pennsylvania.

Even though cable television is not yet forty years old, it has already spawned fables. The story of its humble beginnings has been told so often, in so many versions, and ascribed to so many different communities, it's hard to tell truth from tall tale. Now that the industry has become large enough to suffer booms

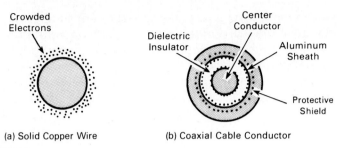

FIGURE 9-10
How Coaxial Cable Overcomes the Skin Effect.

and busts and to warrant consideration by Congress, it is well to appreciate that the early visions for cable were certainly not any that might challenge the nation's established broadcast industry.

When broadcast television became a reality for the nation in the late 1940s many areas could not receive any signal at all or, at best, only a poor quality signal from but one television station. These areas were shielded from the line-of-sight television transmissions by either natural barriers such as mountains or by manmade barriers such as the skyscrapers on Manhattan. One such area was Lansford, Pennsylvania, where the mountains shielded residents from television stations in Philadelphia, only sixty-five miles away. Not surprisingly, radio and television sales were hardly booming in an area where but one poor signal was available. So an enterprising salesman, Robert Tarleton, climbed to the top of the mountain and installed some simple antennae to service those who had had the courage to purchase television sets. New purchasers wanted their antennae placed on the mountain so they, too, could receive signals not only from Philadelphia but from New York City and Baltimore. Soon, the mountaintop was a forest of antennae with bundles of cables cascading down the mountainside, hardly scenic and not terribly reliable. However, the sales of television sets boomed and Tarleton soon convinced some local businesspeople that there was money to be made from delivering wired television to the community. He built a *master antenna* on the mountain, a single structure on which several antennae were mounted, which received signals from the three cities. He also installed amplifiers that "cleaned up" and enhanced the weak signals and wired the television households wanting his service with coaxial cables strung on telephone poles. Tarleton's Panther Valley Television system (Figure 9-11) charged $130 for installation and $3 per month for service.

Thirty-six years later, the basic principles of cable television design have changed very little. There are still the master antenna with associated amplifiers and other equipment to sharpen up the local and distant signals, coaxial cables that run over or under city streets, and a cable that drops the signal into the households. Thirty-six years later, in 1985, 43 million, almost half the total television households in the nation, subscribe to cable television.

The Cable Spectrum That television can be delivered on a wire is interesting but it is not what makes cable the institution-shattering technology it is. The significance of cable television is that it frees us from the constraints of the over-the-air radio spectrum. With the radio spectrum essentially "encapsulated" in the coaxial cable, we can make our own rules on how we choose to allocate frequency regions.

This significant breakthrough can best be understood by comparing the cable spectrum to the radio spectrum described earlier in this chapter. In Figure 9-12 we show the radio spectrum of Table 9-2 and the cable spectrum as we have decided to construct it for cable television.

As noted earlier, over-the-air broadcasts take place in three regions of the radio spectrum:

- Television channels 2 through 6 at 54MHz to 88MHz,
- Television channels 7 through 13 at 174MHz to 216MHz, and
- Television channels 14 through 69 at 470MHz to 806MHz.

To ensure that channel recognition would be maintained wherever possible, cable channel numbers are made identical to the over-the-air channel designations. You will note, then, that the VHF channels 2 through 13 are assigned in the cable spectrum precisely where they are assigned in the radio spectrum.

FIGURE 9-11
Panther Valley's CATV System.

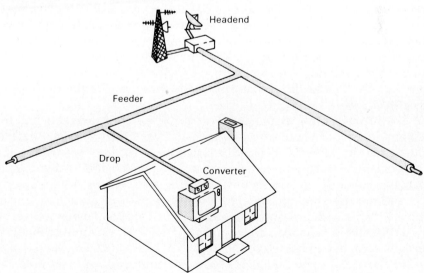

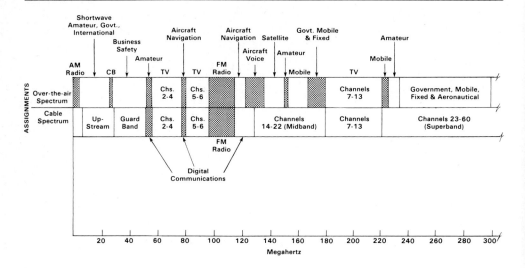

FIGURE 9-12
The Over-the-Air Cable Spectra: Cable Structures the Radio Spectrum to Its Needs.

However, to expand cable's coverage to the UHF channels 14 through 69 it is necessary to expand the range of the cable to 806MHz. We found that it is very difficult and expensive to design amplifiers that must deliver linear, that is, high-fidelity amplification, over a very wide range of frequencies. (Refer to our discussion of radio receiver technology and in particular to the reason for the invention of the superheterodyne receiver.) To make matters even more difficult for designers, cable television amplifiers must operate reliably when exposed to the changing temperature and humidity conditions found on poles or in underground ducts. Consequently, early cable systems were generally limited in their bandwidth, usually just enough to provide the twelve VHF television channels.

To provide television channels 2 through 13 required the cable and the amplifiers to have a bandwidth of between 54MHz and 216MHz. But the frequencies from 121MHz and 174MHz are not even used! Over-the-air, this range is occupied by a number of users, including police, fire, amateur radio, aircraft, and military communications. Since the cable delivers its signals along a wire and not over-the-air, there is no reason why these same frequencies cannot be shared without danger of interference. Cable could now deliver all of the VHF television channels and a portion of the UHF television channels in the frequency range of 54MHz to 216MHz.

In the 1970s the semiconductor chip was introduced into the design of analog amplifiers in the form of integrated circuits. These *hybrid amplifiers*

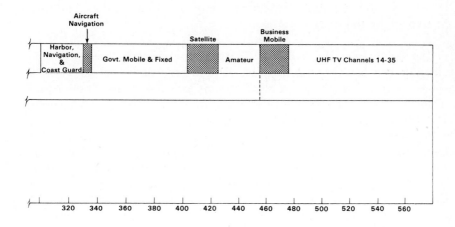

FIGURE 9-12 (Continued)

were capable of broadband operation—that is, they could reliably and consistently amplify linearly over a wide range of frequencies, not subject to temperature and humidity variations. Broadband amplifiers capable of working over the range of frequencies from 10MHz to 450MHz were developed and there is no doubt that even higher and broader bandwidths will be available in the future, perhaps clear up to 890MHz.

The cable spectrum, however, need not parallel the radio spectrum. Because the signals are "encapsulated" by the cable, and not released over-the-air, cable can share frequencies among different uses. Thus, cable television channels 23 to 36 are shared with government-fixed and mobile communications and with another band of aeronautical radio. Cable television channels 37 through 60 are shared with harbor, navigational, and Coast Guard users, as well as satellite communications. Furthermore, a cable television system can reuse its own frequencies by installing more than one cable; many communities in the nation today boast systems with 120 channel capacity and with even a third cable for sending signals "upstream" from the receiver to the sender.

Cable television's challenge to broadcasting is based upon this ability to manage the spectrum in ways impossible for over-the-air broadcasting. A wired system immune from radio interference and with growing channel capacity can perform numerous functions either difficult or impossible for traditional broadcasting. These include the direct and immediate importation (in "real-time") of distant television signals, signals from one part of the nation to another, the origination of signals that will reach into neighborhoods rather than entire cities (narrowcasting), and the provision of pay-television. Two-way interactive services are also possible and this raises the question of how cable television and the telephone will get along with one another. Let's examine just how they will work together.

The Cable Television System The structure of cable television today has changed remarkably little from that of Panther Valley Television in 1949. A cable television system consists of three major subsystems: the headend, the distribution system, and the subscriber drop. These same components graced Tarleton's Panther Valley Television system.

All signals begin their journey to the subscriber from the *headend*. At the headend antennae capture local broadcast signals—signals that have been sent over long distances by microwave relay—and television receive-only earth stations for signals that are generated elsewhere and are delivered to cable system by satellite for retransmission to subscribers. These include the pay-television programming that has become the mainstay of the cable television industry. The headend also receives for retransmission signals that are locally originated by the cable company including the public benefit channels (public access, education, and local government channels) often required by the franchising authorities.

The signals are processed, noise eliminated, and information amplified, enhanced, and made ready for distribution over the long cable runs. Video character generators—computers that accept digital information from keyboards or over the telephone and convert it into video displays—deliver time and weather, news, sports scores, program listings, stock market quotations, and other information. Headend technology also has evolved a considerable data processing capability that can be used for high-speed data transmission. Remember, cable is a broadband transmission technology and very suitable for high-speed data transmission. Consequently, cable television is often seen as a technology for by-passing the local loop of the telephone system, especially for private-line, high-speed data or video services, a prospect that has become increasingly attractive to business since the restructuring of the nation's telephone industry.

The cable television *distribution system* is the tree topology we described in Chapter 5, consisting of trunks and branches, and is quite different from the telephone's star topology. This has important consequences for cable's ability to serve as a telephone by-pass technology. In a tree structure, all of the signals processed by the headend for delivery to subscribers are sent down the trunks, to distribution and feeder cables and to the subscriber drops into every household. All cables in the system are required to have the bandwidth and the amplifiers capable of transmitting all of the signals. Broadband hybrid amplifiers are spaced every mile or so along the distribution trunks and feeder cables. The cost per mile of cable, an important parameter that must be compared to the number of potential subscribers in that mile, is determined to a considerable extent by the number of amplifiers required to deliver a suitable signal to the television household.

The number of channels corresponds to the number of signals the cable television operator chooses to offer subscribers. Channel selection is made by subscribers by means of channel selector switches or converters in their homes. Because all of the signals are always present in the cable, it is relatively easy to pirate signals that may not be authorized, such as the scrambled signals that are provided to pay-television subscribers. Those who pay have devices within their converters that can reassemble the original signal.

Now think back on the telephone architecture in which only one channel at a time as requested is delivered to a subscriber and there are as many channels as there are subscribers, albeit of much lower bandwidth (4KHz as compared to 450MHz on the cable). Switching is performed at central offices and it is relatively difficult, although not impossible, to pirate or listen in on other people's conversations.

In our discussion of the cable spectrum, we noted that cable's ability to structure the radio spectrum according to its own needs without fear of interference is the key to its important role in modern broadcasting. Figure 9-12 shows a portion of the cable spectrum in the 5MHz to 35MHz range that is set aside for upstream signal transmission for two-way services. These services could include data transmission which will be switched at the headend for delivery downstream on channels set aside for data transmission. For example, banks wishing to transmit high-speed data from one branch to another could use these special services. Another use for the upstream capability is video conferencing, however, this would require the entire channel.

Early experiments with providing interactive services on the cable resulted in some interchannel interference between the digital and video signals. Consequently, a second or even a third cable usually is installed to pass by only those areas in the city being cabled that are likely to require interactive data or video services, such as industrial parks, streets in neighborhoods zoned for business, schools and college campuses, and government centers. This cable has become known as the *institutional cable*.

Finally, there is the *subscriber drop* which delivers cable television signals into the subscriber household and which includes the necessary equipment for making selections from all of the channels delivered. Much of the intelligence of the cable system is in the subscriber equipment. A growing number of services are available in packages called service tiers. These tiers range from the very basic package which delivers only the local television signals to tiers that deliver in addition to the local services and imported signals, several pay-television channels. Since not every subscriber will purchase the identical mix of services, it is necessary to provide the means for selecting among these mixes so that those who purchase will be billed for the appropriate services and those who do not will be denied access to them.

The simplest converters are devices that will provide the subscriber with more than twelve channels of television, by converting the ultra high frequency television channels 14 and up for delivery in the midband range shown in Figure 9-12. More complex and expensive converters are required to unscramble the scrambled pay-television signals transmitted down the cable. Scrambling is usually performed by altering the transmitted pulses required to make the picture hold still on the screen; the unscrambling device simply provides that signal.

As we shall see in Chapter 13 on network information services, cable television is viewed as an attractive means for delivering banking, shopping, and other interactive customer services directly into the home. This requires an even more complex consumer device, one that can generate intelligence in the form of digital pulses for transmittal upstream. Since the number of converters required equals the number of subscribers, the cost to the cable operator can be quite high, especially if the operator wishes to encourage subscribers to purchase interactive services by providing these converters either at no charge or for lease.

CABLE TELEVISION AND THE SECOND ERA OF TELEVISION

In the mid 1970s cable television was seen as the harbinger of a second era of television, not only in the United States but in many European nations. Cable would not only provide television services where there were few or no signals at all but also would deliver more television diversity because of its almost unlimited number of channels. Furthermore, cable television would do what over-the-air broadcasting had been unable to do—meet the broadcasting needs of communities, even neighborhoods. Cable was to become the second major wired telecommunications system and could deliver services telephone carriers could not or would not deliver.

The long, often confusing, and certainly contradictory legislative and regulatory histories of cable show that Congress and the FCC did not have a clear view of what cable was or should become. It was not long before cable was seen by the broadcasters as a competitor, for by importing distant signals into a community and by locally originating programs, cable fragmented the television audience. A fragmented television audience could reduce advertising revenue to a single channel. Cable television also was seen as a threat to the motion picture industry and especially to theater owners, for pay television in the home would reduce theater attendance. An already beleaguered telephone carrier industry saw the possibility that cable could skim off its highly profitable data transmission and video businesses.

The FCC saw cable as "ancillary to broadcasting" and therefore imposed regulations. Cities required cable television firms to obtain franchises since they were using the city streets. Some states imposed their own regulations on cable. Cable television was mired in a complex three-level regulatory structure. Nevertheless, when cable showed that it could add to the revenues of the motion picture industry and that the demographics of movie going would not be drastically altered, pay-television became the spur that moved the industry rapidly to a subscriber penetration of almost 50 percent of all television households in the nation in 1985.

Now cable television itself is under challenge by what many call "wireless cable" created by the more efficient management of the radio spectrum.

Wireless Cable

As early as 1947, broadcasters had sought the means to provide for-pay television over-the-air. In 1950 WOR-TV tested a scrambling system called Skiatron and in Los Angeles KTLA tested a similar system called Telemeter. By scrambling the television signal much as the cable television folks do today, they proposed to deliver movies and other entertainment only to those subscribers who purchased a decoder. Ordering up a program was accomplished by telephone, thereby ensuring that the caller would be billed for the entertainment purchased. Some schemes for ensuring payment and overcoming the relatively easy piracy were to require subscribers to put coins in their decoders in order to set the unscrambler working. The audience was there, but the technology was not!

In the 1960s the FCC recognized that there was a demand for pay-television over-the-air as well as on the cable. The Commission invited applications from commercial television stations that might want to explore the prospects for *subscription television* (STV). To overcome or appease the strong opposition of the theater owners and the broadcasters, the FCC limited these trials to no more than three years and to cities that had at least four commercial television stations. No more than one STV operation was allowed in any one community and the broadcast stations had to continue to provide unscrambled operations for at least twenty-eight hours per week.

Not too many applicants came forward; the investment in hardware and marketing seemed too high given the three-year limit.

It wasn't until 1977, long after cable operators had shown that pay-television was a profitable and promising business and the FCC had rescinded a host of restraints on the kinds of films that could be shown, that commercial operations began in earnest. As might be expected, these operations were in cities that were not yet cabled and the audience response was encouraging, to say the least. In a section of Los Angeles that had not yet been franchised, more than 125,000 households in an area of about 300,000 television households became subscribers. For about $22 per month, a household could receive movies and sporting events not available on traditional television; so, unless prospective subscribers were in poor reception areas, they really did not need cable.

STV subscribers turned out to very loyal. Cable's multichannel offerings of sports, 24-hour news, several schedules of for-pay movies, and several superstations, importing television from distant cities, attracted the single-channel STV subscriber to multichannel cable but often only after several years of hard selling.

Cable and the broadcasters were still not home free!

Among the many bands allocated for television was the band from 2500MHz to 2690MHz for Instruction Television Fixed Services (ITFS), a service promoted by the educational establishment for the delivery of educational and instructional programming. In every city, four groups of frequencies were set aside but after almost thirty years, little more then 35 percent were actually applied for and fewer were in active use. The major users continue to be the Catholic Church, especially the Archdioceses of San Francisco, New York, and Philadelphia, and several public school districts, notably in San Diego and Spokane. In general, the ITFS frequencies are underutilized.

In the early 1970s, the FCC authorized a portion of the ITFS spectrum for commercial uses, specifically for the transmission of data and other services that could operate within a 2MHz bandwidth. The characteristics of the new *multipoint distribution services* (MDS) were quite similar to those of ITFS—a point-to-point range of about twenty-five miles. While the complaints from the educational community were loud, poor usage of the ITFS channels did not support these claims and, in any case, there was still video spectrum available for educational uses. Nevertheless, the FCC set a precedent for further reallocation of spectrum in this region.

Improvements in amplifier design, to a great extent due to the marriage of analog and digital technology in the hybrid amplifier, led to the development of video broadcast systems that could operate at ever higher frequencies. Meanwhile, subscription television and cable television were showing that an audience did exist for television for pay over-the-air.

Soon the traditional broadcasters had three competitors: cable television, subscription television, and multipoint distribution services. The broadcasters and theater owners had made a sort of peace with cable, they felt no threat from an already diminishing STV business, and assumed a single channel MDS would very likely suffer a similar fate in the long run. However, the FCC continued to divide up the ITFS spectrum and in response to requests from MDS operators expanded their portion to allow *multichannel multipoint distribution services* (MMDS).

There is some evidence that the average cable television subscriber watches about eight channels of television, at least five over-the-air offerings and three special cable offerings including pay-TV. If this is indeed so, then MMDS may be a low-cost competitor to cable—wireless cable, if you will, for with MMDS there is no need to string wire on poles or dig up city streets to run cable to deliver up to five channels of pay-TV, sports, and a superstation or two.

The Satellite and the New Television

Developments in the design of television receive-only earth stations, the TVROs we discussed in Chapter 4, have followed the trends we have been observing in the broadcast technologies. Hybrid designs that marry digital microprocessors to analog broadcasting have dramatically reduced the cost of the low-noise amplifiers and, consequently, the cost of the receive-only earth stations. With almost all of the cable services that attract subscribers now delivered via satellite, it is no wonder that individuals are thinking seriously about having their own TVROs and by-passing the cable entirely. Throughout the United States, satellite "dishes" are popping up on lawns and in yards like mushrooms. In the mid-1980s a TVRO with motor to drive the dish between several "birds" could be purchased and installed for less than $4000.

Apartment blocks and condominiums have found that even with the attractive discounts cable operators might offer them, they are better off installing a TVRO and delivering services directly to their tenants. In many cases, these buildings already have master antennae that pick up and distribute the local television signals to their residents. *Satellite master antenna television* (SMATV) has emerged as yet another competitor to cable. Indeed, with the increasing popularity of multifamily dwellings brought on by high interest rates and the high cost of single-family homes, more and more television households in a cable television franchise area are likely to be multifamily. *SMATV* systems could siphon off a significant number of those very households that would likely subscribe to cable television.

Finally, we must mention, once again, the direct broadcasting satellite (see Chapter 4). When the DBS was first proposed in the early 1980s, it was believed that its primary audience would be in those areas of the nation not likely to be wired, estimated to be about 15 percent of all television households. It was expected that the cost of the receive-only terminals would fall to the $300 level until well into the next decade. But experience with SMATV, the surprising number of households that are investing in TVROs at $2500 to $5000 each, and the rising cost of monthly cable subscriptions (cable operators claim that they cannot make a reasonable return on their investments until the average monthly rate reaches about $60/month or $720/year), we ought not to underestimate the impact the DBS may have on our broadcast infrastructure.

Just what services direct broadcast satellites are going to offer is quite unclear, but it is those services, not technology that are being marketed. That subscribers are searching for new television is evidenced by the 40 percent "churn" rate often experienced by cable television operators. Churn can be interpreted as a measure of subscriber dissatisfaction since it is the percentage of disconnects experienced by cable operators. Disconnects can be due to the mobility of today's households, but even with a rate of mobility in the neighborhood of 15-20 percent, a 40 percent rate of churn means that two out of every ten subscribers disconnects each month because they are dissatisfied with the service.

THE MANY ROLES FOR BROADCASTING

There is no better example than broadcasting with which to illustrate the degree to which culture and technology interact. Technology responds to cultural imperatives and then turns around and alters traditional institutions which, in turn, affects a society's cultural norms. For that reason alone, this is a good time to explore the many roles of broadcasting.

In the United States it is difficult to identify a "decision" to introduce broadcasting. As with so many of our consumer products, when a particular market looked promising entrepreneurs sought capital as well as legal authority to establish a broadcast business. In the early 1900s, Congress recognized the need for an orderly process by which entrepreneurs could share the potentially profitable broadcast spectrum with such public uses as ship-to-shore radio. The Wireless Ship Act of 1910 required all ships with more than fifty passengers to carry Marconi's relatively new invention. Unfortunately, the act itself did not require those receiving a message to act with alacrity, as was shown in the 1912 sinking of the Titanic. Consequently, Congress enacted the Radio Act of 1912 which not only required ships to carry a "wireless" but also required everyone operating on the radio waves to obtain a license from the secretary of commerce and be accountable to Congress for how that license was used.

While the secretary of commerce could not deny a license and could only assign particular frequencies or wavelengths to applicants, the Radio Act did set a pattern for government involvement in the radio business that culminated in the Communication Act of 1934. This act recognized the scarcity of the broadcast spectrum and, since not everyone wanting a broadcast license could get one, established certain requirements broadcasters must meet in order to ensure that broadcasting is "in the public interest, convenience, and necessity," a phrase borrowed from the railroads and which first appeared in the Radio Act of 1927. Congress carefully avoided any statement of what broadcasters were expected to say, for that would have infringed upon their First Amendment rights. For the United States, the role of broadcasting is often seen as strengthening the First Amendment, giving added weight to the Jeffersonian ideal that an informed populace sampling from a rich marketplace of ideas is the best insurance for the preservation of a democracy. Further, Congress made it quite clear that this marketplace was to be occupied by private entrepreneurs competing for the attentions of the public.

Today, there are more than 9400 radio stations and more than 1100 television stations in the nation. The heterogeneous nature of American society is reflected in the allocation of radio and television licenses to communities in order to ensure a high degree of localism. Our goal was program diversity, for opinions, information needs, and tastes varied as widely as did the multicultural, multi-ethnic, and multieconomic population of a growing country.

Broadcasting is a multibillion dollar industry in which the station broadcaster is but the final link in a massive interconnected chain of participants that includes advertisers, creative artists, producers, promoters, directors, stagehands, equipment manufacturers, maintenance and repair personnel, and bankers and brokers. All of these talents are geared to the single task of capturing the audience's attention, ensuring that the most eyes and ears are glued to a particular television channel or radio station.

It becomes quite clear, then, that should other means be found to capture the attention of the listeners and viewers, the stations now "on the air" would lose their audiences, and with fewer people watching and listening, the advertiser would not have to pay as much for delivery of the message. The new broadcast and narrowcast technologies have taken their toll on audience loyalty and have struck fear in the hearts of broadcasters while at the same time raising viewer and listener expectations.

There is little doubt that the nature of the broadcasting industries, and specifically the television industry, is rapidly changing in the United States. But is the nature of television changing?

Almost twenty years ago, when the promise of cable television appeared bright, there were some who forecast that the future of television is radio. Radio had not succumbed to television just as television did not kill the movies. Indeed, radio has become richer because of the competition among the many AM and FM stations in communities around the country. Furthermore, since the cost of radio operations had been falling as rapidly as computer-automated studio operations were introduced, large audiences were not necessary to keep radio afloat, thus creative programming geared to the special needs of smaller audiences became possible.

Will the future of television be radio? The jury is not yet in; we haven't had enough experience with multichannel viewing to know if the industry will recognize that change is not only necessary but might well be profitable. Research to date points to the discouraging finding that the availability of more television channels does not necessarily mean more diversity. But this same research does show that cable operators are responsive to the special needs of their communities. Is this because they are responsible broadcasters responding to both market and social needs or because they are regulated by local governments who can decide whether or not they will get their next rate increase?

Can it be left to the marketplace to decide how broadcasting should serve the public? Is there enough competition between entertainment and information providers, now that we have learned to manage the radio spectrum in new ways? Will the promise of broadcasting live up to what the technology can deliver without some form of government intervention?

The turmoil over television's role is not unique to the United States. Throughout most of the world, broadcasting is government owned and operated, the exceptions being Canada and Great Britain where there are both government and privately owned broadcast systems. Thus in many countries, broadcasting is expected to serve the state. For example, in France, broadcasting seeks to "preserve, enhance, and glorify" the French language and culture and broadcasters are given credit and funds partly on the basis of how well they serve children. In West Germany and Great Britain, high-quality drama and music are programmed during prime-time hours to ensure that high standards of appreciation are created and maintained. In Sweden and the Netherlands, great care is taken to serve minority groups, and in Indonesia, the major task of the broadcast system is to link the more than 145 million inhabitants spread across 13,677 islands with a national identity and by a common language.

Broadcasting in the service of the nation requires that the systems be supported entirely by taxes or license fees paid for the privilege of owning a radio and television receiver. In recent years these fees and taxes have risen sharply as the cost of programming has risen in response to appreciation of public tastes. Television is a medium that seems to create high standards of visual "literacy" and even a child soon recognizes quality program production values. In the United States, this rapid increase in the cost of television production resulted in the creation of the major broadcast networks; the revenues from nationwide advertising were required to produce the quality programming the public demanded despite a reduction in program diversity the local distribution of television licenses had sought to achieve. In Europe, Japan, and other nations where the operating costs of television are provided through taxes and license fees, the unpopularity of yet another rate increase has resulted in a serious loss of operating revenues. For example, the French and Italian motion picture industries, so well-received throughout the world, will not provide programming to their state television networks because of the low payments they receive for their work.

In these nations as in other countries, the new broadcast technologies provide opportunities for new sources of revenue through the delivery of additional channels or services. With this development, has come an alteration in national policies for broadcasting recognizing that while cable television will create new revenues, it also may lead to the loss of control of programming by the state. The technology of wired radio and television creates the opportunity for locally originated productions, productions that cannot be controlled by the ministries of information in the nations' capitals.

As we noted previously, the technology of broadcasting is universal, but how it is used varies from nation to nation. New technologies which allow access to new bands in the electromagnetic spectrum could very well alter traditional broadcast institutions. In the United States, the new technologies have attracted new entrepreneurs who segment the audience by providing more diversity and thereby enriching the marketplace of ideas. In those nations with government-controlled broadcast systems, the new technologies are segmenting the audiences by offering more diversity in order to create more revenue and in so doing, challenging the government monopolies in broadcasting. It is most interesting to speculate about how these nations will adjust to a rich and often politically threatening marketplace of ideas where none existed before. In Chapter 4 we noted that some have called the satellite the great hope for democratizing the world. Perhaps earth-based broadcasting should be added to our menu of world democratization tools.

ADDITIONAL READINGS

Writing about broadcasting, especially television, has been a popular activity of journalists and scholars; academic careers have been built on commentaries about television and its impact on life in our time. Because broadcasting serves so many interests, the literature spans a wide range from technology to politics, from creative practices to how radio and television assists in national development. Choosing from this vast array is no mean task; however, we have sought to select what we believe are among the best readings for several of the interests broadcasting serves.

Broadcast, Cable, and the New Technologies

Baldwin, Thomas F. and D. Stevens McVoy, *Cable Communications* (Englewood Cliffs, NJ: Prentice-Hall, Inc., 1983).

Gross, Lynne Schafer, *Telecommunications: An Introduction to Radio, Television, and the Developing Media* (Dubuque, IA: Wm. C. Brown Company Publishers, 1983). An excellent view of the broadcast industry.

Mahoney, Sheila, Robert Stengel, and Nick de Martino, *Keeping PACE With the New Television: Public Television & Changing Technologies* (New York: VNU Books, International, 1980). For an overview of the new television technologies see pp. 47–224.

Miller, Gary M., *Handbook of Electronic Communication* (Englewood Cliffs, NJ: Prentice-Hall, Inc., 1979), pp. 100–137 for an excellent discussion of the technology of single-sideband communications and pp. 266–313 for an excellent review of the television.

Stafford, R. H., *Digital Television: Bandwidth Reduction and Communication Aspects* (New York: John Wiley & Sons, 1980). For the adventurous, an excellent view of the likely future of television technology.

Wurtzel, Alan, *Television Production,* 2nd edition (New York: McGraw-Hill, 1983). See pp. 23–54 for an excellent discussion of the technology of the television camera. The standard book for those interested in television production, too.

The Business of Broadcasting

Barnouw, Erik, *Tube of Plenty: The Evolution of American Television* (London: Oxford University Press, 1977).

Brown, Les, *The Business Behind the Box* (New York: Harcourt Brace Javanovitch, 1971).

Mayer, Martin, *About Television* (New York: Harper & Row, 1972).

Noll, Roger, Merton Peck, and John J. McGowan, *Economic Aspects of Television Regulation* (Washington, DC: The Brookings Institution, 1973).

Broadcasting and Society

Comstock, George, *Television in America* (Beverly Hills: Sage Publications, 1980). Provides an excellent summary of the Surgeon General's findings.

Katz, Elihu, and George Wedell, *Broadcasting in the Third World: Promise and Performance* (Cambridge, MA: Harvard University Press, 1977).

Lang, Gladys, and Kurt Lang, *The Battle for Public Opinion: The President and the Press and the Polls During Watergate* (New York: Columbia University Press, 1983).

Surveying Broadcasting

Head, Sydney W., and Christopher H. Sterling, *Broadcasting in America: A Survey of Television and the New Technologies,* 4th edition (Boston: Houghton Mifflin Company, 1984). The standard history brought up to date.

Sterling, Christopher H., *Electronic Media: A Guide to Trends in Broadcasting and Newer Technologies, 1920–1983* (New York: Praeger Publishers, 1983). This is a single volume reference guide with useful statistics about trends.

CHAPTER **10**

MOBILE RADIO: THE FASTEST GROWING TELECOMMUNICATIONS INDUSTRY

We are a mobile society; mobility is imbedded in the American psyche.

The very first colonists settled in Virginia and New England only temporarily and quickly sought to move as far away from their neighbors as possible. As rapidly as their horses could carry them, their wagons survive the rough terrain, and Indians be pushed aside, the new Americans moved from one coast to the other. The railroad, the horseless carriage, and the airplane were the means by which we expressed this apparently unique American desire for mobility.

The great historian of the west, Frederick Jackson Turner, viewed the history of the United States in terms of this inexorable westward movement and speculated that once the nation had been traversed we would no longer have the energy and drive of our early years. Still, there was Hawaii to explore and Alaska to settle and when we had done so, we moved about within our fifty states. Americans are, indeed, the most mobile of peoples. Census figures show that at least 50 percent of the American population moves every five years and of these, about 20 percent move to another state. The rootlessness and transience of our people have been studied by sociologists and demographers, all seeking to find some way of interpreting what appears to be the breakdown of family and community.

ELECTRONIC MOBILITY

When we had succeeded in exhausting our geographical mobility, we sought the mobility that electronic communications could offer. Communicating at a distance, by way of the telegraph and telephone, became the new frontier for expansion. As noted in Chapter 6, contrary to what many expected, the telephone may have actually increased our opportunities for mobility rather than reduced them. The Third Wave will not likely see mobile Americans "at home." Rather our society will use modern telecommunications as a further excuse for more travel. Links to family and community will be maintained via telecommunications. There will be even fewer reasons to be stationary since communicating on the move, while driving or while traveling in the air and on the sea, provides the connections we desire with family and community. Indeed, Daniel Bell notes that the word community will take on new meaning, no longer bounded by geography but rather encouraged by common interests and promoted by telecommunication networks.

Mobile communications—mobile radio and telephone communications—the subject of this chapter, are examples of how telecommunications increases mobility rather than decreases travel.

TOUCHING SOMEONE ELECTRONICALLY

Mobile telecommunications is not a new phenomena. Not long after Marconi sent and received his first wireless telegraph message in 1895, we bridged the gap between ships at sea and the telegraph networks on land and communicated with remote locations not yet served by the telegraph using primitive but effective radio transmissions. In 1899 a lightship off the coast of Fire Island in New York was linked by radio to telegraph systems on the mainland; radio communication with a floating balloon was achieved in 1908. While the balloon was not moving very quickly this was, certainly, an early example of communicating on the move.

By 1929, a high-seas public radio telephone service was initiated between the ship Leviathan and the U.S. mainland and every Navy vessel was soon equipped with mobile radio systems with which to keep in touch with U.S. ports. There are some wonderful stories in Navy lore about how surgery on board ship was performed by corpsmen following instructions from doctors in mainland hospitals delivered via radio-telephone.

Police patrol cars in New York City were equipped with radios as early as 1916. And the notion of a mobile patrol in constant communication with other patrols and with a control center emerged in New York City and in Detroit in the late 1920s. Amidst the confusion of early broadcast radio in 1927, federal regulators searched for some order as amateur and commercial radio enthusiasts moved in on mobile radio; in fact, we can safely say that this was the beginning of the commercial mobile telecommunications industry. Today it is the fastest-growing telecommunications industry, not only in the United States, but throughout the world.

Why is this so? What can we say about the reasons for the growing demand for mobile telecommunications?

In general, societies characterized by long trips between home and place of work seem to generate demands for mobile services. Societies where there is a great deal of intercity travel and dependence on trucks and railroad for the transport of goods also create a demand for mobile services. Where there are public service people and goods delivery networks, such as taxis, buses, trucks, aircraft, and railroads benefiting from communications with a central location as well as with each other in order to work efficiently, mobile communications are in great demand. Finally, services to be delivered to homes and offices—documents and other special written or printed information, repair, safety, and emergency assistance—all create a demand for mobile telecommunications.

Interest in and demand for mobile telecommunications are not generated solely by economic, safety, and efficiency needs; socializing on the citizen's band, the CB media event of the '70s, provided some evidence that the desire for geographic mobility matches the desire to maintain contact, to keep in touch while on the move. CB radios weren't purchased solely by truck drivers emulating Smokey and the Bandit; "Big Mommas" have included a former First Lady.

Some few studies that monitored CB conversations showed that keeping track of the police on highways was not the most popular topic of conversation. Making appointments for subsequent face-to-face meetings led the list followed by keeping in touch with the community of CB users, talking amongst themselves about their hardware much as personal computer users do today on their networks, comparing hobbies and leisure activities, and, of course, complaining about the weather. By far the majority of the messages people on the run transmit to other people on the run deal with what sociologists call expressive information: gossip, personal feelings, socializing on the electronic network, in short, not much different from many of our telephone conversations.

While the appreciation and even the benefits of mobile communications transcend pure efficiency and economics, they may indeed include a variety of intrinsic or expressive benefits—the purchaser initially wants to save time, be reached, and reach others, in other words communicate efficiently and easily. Reliable, noise-free and error-free communications everywhere in the user's travel geography, minimal waiting time for a "line," and security and privacy are the services for which subscribers to mobile services are willing to pay.

MAKING A LITTLE GO FAR

In our discussion of spectrum in Chapter 9 we noted that the Federal Communications Commission has assigned frequencies for different kinds of radio services: from about 800KHz to about 1400KHz for AM Radio, VHF TV and FM radio in the space around 100MHz, satellites in the 4–6GHz range and on up. Until recently, mobile radio communications were confined to very narrow bands scattered throughout the radio spectrum wherever they could be squeezed in—30–50MHz, 150–174MHz, 450–5121MHz all shared with other services, an arrangement not at all suited to the very rapidly growing demand for mobile communications. With more and more users essentially on the same line, long waits for "dial tones" or channels and interference were everywhere. The FCC added another range of frequencies in the 1960s—450–470MHz—devoted solely to mobile users, but the waiting lists continued to grow and the service to deteriorate. Users placed their calls through an operator who managed up to forty subscribers on a single line—a very large party line—and, as you can imagine, tempers often flared as users lined up for service.

It was clear that demand was far outstripping supply. As Figure 10-1 shows, mobile communications grows in parallel with the U.S. population and motor vehicle growth. Even conservative projections showed the FCC the need to allocate more bandwidth for mobile radio. Clever ways of using the spectrum, some of which we have already discussed and others that we will shortly, only seemed to attract more users to the crowded spectrum, a wonderful example of the highway or freeway effect every Californian knows. (No matter how many lanes you plan for they are filled years sooner than your projections said they would be.)

Under great pressure from AT&T, a major provider of mobile radio/telephone services, the FCC made available a large block of frequencies in the UHF band (see Table 10-1) between 806 and 947MHz (this has become known as the 900MHz band). The demand for mobile services nevertheless continues to grow, faster than the supply of bandwidth, and, consequently, channels are

FIGURE 10-1
Charting the Growth of Mobile Radio. (From James Martin, *Future Developments in Telecommunications*, 2nd ed. Englewood Cliffs, N.J.: Prentice-Hall, 1977. Used with permission.)

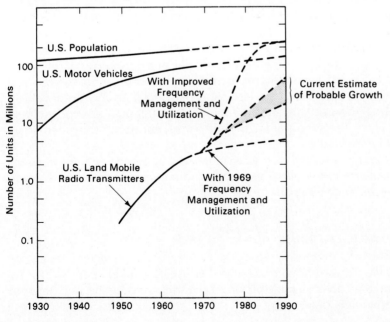

TABLE 10-1
SPECTRUM ALLOCATION FOR MOBILE RADIO

Transmission Media	Application	Frequency	Wavelength
Microwave Radio ↕	Satellite to Satellite Microwave Relay Earth to Satellite	10GHz	1 centimeter
	Radar	1GHz	
Shortwave Radio ↕	UHF TV		
	Mobile Aeronautical		
	VHF TV & FM Radio	100MHz	
	Mobile Radio		10 meters
	Business Amateur Radio International Citizen's Band	10MHz	
Longwave Radio ↕			100 meters
	AM Broadcasting	1MHz	

crowded, and often noisy. Tempers, while somewhat cooled down, still often reach the boiling point.

The increasing demand for mobile services is not solely a U.S. phenomenon. Indeed, in almost all the developed and even developing nations of the world there seems to be a dramatic increase in the demand for mobile radio and telephone services, an increase that also parallels population and motor vehicle growth. As regulatory agencies and governments throughout the world increase the number of channels assigned to mobile communications, they seem to be filled almost immediately.

How can these increasingly crowded pathways for mobile communications be used to make room for all of the mobile communications now on the road and for those waiting to get out on the highway?

HOW TO CROWD ELECTRONIC HIGHWAYS AND GET AWAY WITH IT

In Chapter 2 we showed how to make better use of valuable bandwidth: by frequency and time division multiplexing a channel, thereby sharing the same channel among many different users.

Instead of assigning channels or frequencies on a fixed basis, we can allocate channels or frequencies as they are needed, allowing the user to search for an unused channel from some group or number of channels until an available one is found. This form of dynamic allocation can be fully automated, especially with the ability to rapidly scan or sample across many frequencies until an unused one is found.

The same frequency can be used many times over if the transmitters and receivers are located in such a way that they do not interfere with one another, as satellites which share frequencies worlds apart.

How the frequencies are used in "real time" also can be managed. This is essentially what telephone operators used to do with patch cord switchboards; they would connect callers only when switch paths and lines were available. These operators have since been replaced by mechanical and computer devices which manage how the limited bandwidth—twisted pairs of copper wires and switch paths, or radio frequencies—are allocated.

As we noted in Chapter 4, satellite communications are reaching into ever higher frequencies on the electromagnetic spectrum to find new bandwidth. Similarly, mobile radio technology can do the same. In recent years, the FCC has allocated additional slots on the spectrum for mobile communications, at ever higher frequencies, and we can expect that as technology improves, additional higher frequencies will be allocated to the roving communicators.

Finally, digital signals rather than analog signals can be transmitted, thereby making the most efficient use of the limited bandwidth (remember pulse code modulation in Chapter 2).

One-Way Paging Systems

The quality of mobile communications services is determined by how well you use the bandwidth in which you must operate. This, in turn, is governed not only by the technology used, but by the messages you send and hope to receive. Consider, for example, paging services. These are the small portable receivers carried on belts or in breast pockets that have almost certainly annoyed you with their beeping in the midst of a concert or a movie followed by a creaking of seats and a rustle of footsteps as someone dashes out of the auditorium. These are the radio paging systems most often used by physicians and plumbers. They provide one-way communications to wandering users, no matter where those users might be within a very large paging area. Depending on the geography, the area reached by paging systems can be fifty to even 100 miles across!

A one-way paging system is shown in Figure 10-2. The answering service or base station may be located remote from the transmitter which is often placed on high ground or on a tall building from which the transmitted signals can reach out into the paging area.

Early paging systems and those still in use in many places throughout the world simply frequency multiplex the assigned channel allocating a small

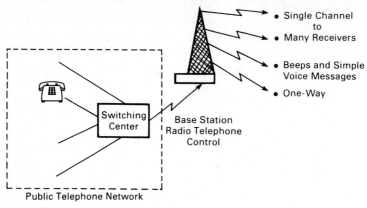
FIGURE 10-2
A One-Way Paging System.

bandwidth to each paging receiver. This is similar to the non-switched communications system we discussed briefly in Chapter 2 (see Figure 2-1). However, simply sending an analog signal on an analog transmission system wastes a great deal of valuable bandwidth.

Today's one-way paging systems send a series of signals, usually five short bursts, from a central base station that activates a tone, a light, or even a silent vibrator in the small portable unit to inform the receiver that someone wishes to communicate with him or her. The paging signals are usually sent twice in order to make certain that the receiver has recognized the message, should there be excessive interference in the area. If the user wishes, the receiving unit can be turned off and a memory component in the unit can store the page for later retrieval when the receiver is turned on again. The person paged uses the plain old telephone to call a prearranged number at the base station which acts as an answering service. The paging operator may have a message for the person paged or may connect him/her directly to the caller.

Paging systems are becoming more automated and easier to access. You can, for example, dial-a-page from a telephone and be connected through the exchange at the base station. The exchange searches a directory for the access code of the paged unit, and sends out the 5-digit code, very much like dialing a number on the telephone.

Paging service providers are usually assigned portions of the mobile spectrum—a fixed bandwidth—and then subdivide this bandwidth by means of frequency multiplexing. Several pagers are assigned to a block of frequencies and are tuned to these frequencies. While all pagers receive the transmitted signal bursts or dial tones, only the pager whose number is being dialed will ring, light up, or vibrate. All that is required in the receivers so that they can respond to the dialed signals are semiconductor switches that can add and compare the dialed number with their own code. We described this semiconductor function in Chapter 3 (see Figures 3-6 (a) and (b)). These are examples of sending a digital signal, the five-coded tone bursts, on an analog carrier, certainly a more efficient means for utilizing valuable bandwidth than simply setting aside a channel for each pager.

A paging signal is sent twice to ensure that it is received. This takes up valuable time and increases the time callers must wait for a dial tone. The time it takes to send multiple pulses and the speed with which these "dial pulses" can be sent determines how many subscribers a paging system can accommodate. Some of today's units can send the five dial pulses in about 200 milliseconds, or two pages in about 400 milliseconds. In an hour the dispatcher can send more than 8500 pages. This can be increased further if allowances are made for a waiting time of as long as five minutes. How we would complain to our local telephone company if we had to wait that long for a dial tone or to contact an operator! Paging, however, is one-way communication with a delayed response, and paging service subscribers appear to be satisfied (or resigned?) to the service delay.

Despite these drawbacks, the demand for paging services continues to grow as our society increasingly becomes a service and information economy. To compete and prosper in this new world requires fast responses to customer demands. The new chip technologies which we discussed in Chapter 3 are improving the design of radio paging systems making dialing faster and providing clever coding devices and other means to increase the number of pagings that can be performed in the limited bandwidth allocated for these services. Today's paging systems can accommodate 200,000 subscribers with the five-minute waiting time. With the rapidly falling costs and size reductions of semiconductors, it may not be too difficult to meet the growing demands of executives, government officials, volunteer firemen, reporters, physicians, plumbers, and others who want to be "on call."

A pleasant human voice is much to be preferred over the sterile "beep" when summoning someone to an emergency whether it be life or property threatening. Consequently, some paging services provide for the transmission of short, usually no more than 15-second voice messages one-way. However, since the voice requires more signal bandwidth, fewer customers can be accommodated.

Because a page is a brief one-way message, paging services can reach out into very large areas, use small, lightweight portable equipment and are low in cost. Paging systems are efficient users of bandwidth having high *spectrum efficiency*—that is, there can be a large number of users served per unit of spectrum.

One-way paging services primarily offer their users the benefits they may derive from always being "on call." Physicians establish reputations and increase their earning power by being available to their patients even when away from their offices. Sales personnel can "keep on top" of sales prospects by being always "on call," and service workers gain new customers by being able to respond quickly to the opportunities that become available when they are on the receiving end of a paging system.

Radio Dispatch Systems

Users of radio dispatch services or the conventional two-way mobile radio are usually in vehicles on the road, on the high seas, or even in the air. They are on an assignment for the delivery of some good or service, on police patrol, performing duties as firefighters, or in a position that requires they communicate with a central location or base so they can be sent on their next assignment. Their next task probably takes them on the road again, often to locations not too distant from the previous assignment. Radio dispatch services are a necessity for taxis, ambulances, police cars, fire trucks, and fleets of delivery vehicles. On large farms, mobile radio systems dispatch tractors from one section to another, saving the fuel required to come back to the barn for instructions and return to the field. Indeed, according to publications coming out of the Soviet Union, utilization of tractors and other agricultural mechanical equipment reportedly increased by almost 25 percent after the installation of a radio dispatch system. We'll have more to say about the economics of mobile communications a bit later.

Most radio dispatch services are simple two-way broadcast systems that operate on a single channel, usually in a *simplex* mode—that is, they have only one voice path available for communications so that signals cannot be transmitted and received at the same time. Hence the "over and out" business. A conventional radio dispatch system is illustrated in Figure 10-3. The dispatcher or control operator is located at the central office where calls for service, or changes in previous instructions are monitored and retransmitted over the single channel network. All of the vehicles receive the message which may be addressed to a single vehicle or to several if the emergency requires, as in the case of police and fire vehicles or ambulances. Privacy is not sought after, indeed, it may not be desirable. Since all of the conversations on the channel can be heard, the dispatcher knows the locations of the vehicles for which he or she is responsible. Indeed, as you have certainly noted, taxis keep the dispatcher informed of their locations in the hopes of picking up nearby fares.

Conventional radio dispatch systems are rather expensive, primarily because relatively few vehicles can be serviced within the bandwidth usually allocated. Within the same channel capacity used for our "beeping" paging systems, a

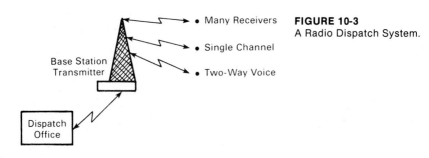

FIGURE 10-3
A Radio Dispatch System.

radio dispatch system can accommodate usually no more than 6000 mobile vehicles but with the chance of getting a busy signal almost two out of three tries! Since all of the users share the same channel, conversations must be brief and the efficiency of the system depends very much on the dispatcher who can control the use of the single channel.

Radio dispatch systems provide another example of sending analog signals on an analog transmission system. And as with paging systems, addressing schemes can be used. Each vehicle can be assigned a digital code, as we did with the paging receivers. Prior to sending a message, the dispatcher "dials" the code. Only the vehicle with that particular code will pick up the message since the other receivers will, in essence, be turned off from the signal and will receive only the hiss of the carrier.

However, this does not increase efficiency; indeed, it reduces utilization of the channel. So while privacy can be achieved, it may not be worth the costs of reduced channel availability.

Multichannel Radio Telephone Services

One way to reduce the probability of getting a busy signal is to provide the radio telephone user more than one channel on which to communicate.

With today's single-channel systems, similar types of users are often assigned the same group of channels—taxis from one company are on the same group, police from one precinct are on the same channel group. Since they are in the same business and cooperation is desirable for the efficient performance of their tasks, being on the same channel group supports their cooperative endeavor. As we have seen, waiting times caused by busy signals can be long and interference often extremely heavy. Furthermore, the operator at the base station must search among the channels in the assigned group to find one that is unused, a process that can be quite troublesome and time consuming.

The marriage of computers and communications, the convergence to which we have been referring throughout this book, is once again in action. Sampling or scanning devices similar to those described in Chapter 2 have been incorporated into the base station control equipment. These semiconductor devices search through the channels assigned to that service and assign the requested call to the first empty one found.

How does the person on the receiving end know what channel is being used for transmission? Remember our discussion of common control switching in Chapter 6? Then you will recall how an additional communication path is added to the switch and to the telephone so that the caller can signal ahead to determine if a switch path is available and the called party phone not busy before actually continuing with the call. This signaling technique has been adopted in the multichannel mobile radio or dispatch system. When a base station finds a free channel, it transmits a single frequency tone over that channel. All of the mobile units sense the tone and lock onto this channel, waiting for the message.

The addition of computer-like functions and control signaling techniques have greatly expanded the capacity of two-way radio systems. Busy signals have been significantly reduced on some of the more advanced systems to as low a probability of only two to five times out of almost 100 tries, especially among the systems used in the dispatch or simplex mode. The higher cost of the advanced multichannel systems (often referred to as MTS or IMTS for improved mobile telephone services) require more users to make the system economical. While this lowers the cost of the mobile receiver/transmitter equipment, the *transceivers,* busy signals are beginning to plague users. Demand continues to increase, especially since the IMTS can interconnect mobile and stationary telephone users. Indeed, you may have difficulty telling the wired telephone services from the mobile telephone services; when you have to wait longer than usual for a dial tone and when the signals seem to fade in and out, only then might you be aware that the party you are talking to is riding about in an automobile, on a boat, or even in the air.

Citizens Band Service

No discussion about telecommunications would be complete without a word about the CB radio. For a short time it made a great deal of headlines and forged, if that is the appropriate word, a unique genre of motion pictures, television films, and popular music. The CB radio gave the trucking industry a special position in our nation's culture, what with numerous "Smokey and the Bandit" films, and rock songs glorifying the intrepid driver fighting his way through hordes of state police with the aid of his trusty CB radio and roadside supporters keeping track of the road ahead. While subsequent investigations tended to show that these adventures with the CB radio were not as frequent as the films and songs would have us believe, the CB radio has found a place in the folklore of the nation.

The citizens band radio is a "party line" for a large number of users (see Figure 10-4). Several classes of service are licensed, all of which are unique to mobile radio in that the allowed power is very low (not more than four watts when used for radio telephony). Consequently, CB radio is short-distance communications, usually for personal use or small business radio communications, but also useful for signaling and remotely controlling model airplanes, boats, and other gadgets with which hobbyists might want to play at a distance. CB radio also can be used for radio telephone services, such as from automobile to automobile or to some fixed location. The limited number of channels available for CB radio as well as the small coverage, between three and twenty miles depending on the geography because of the limited power allowed, severely limits its use as a radio telephone.

In a CB radio system there need not be a base station or controller of the channels; control of the channels is entirely up to the users themselves. However, there must be an operating protocol the users agree to use, otherwise there will be nothing but interference and fistfights. It is a credit to civilized man that such fights over channel interference rarely occur and that many

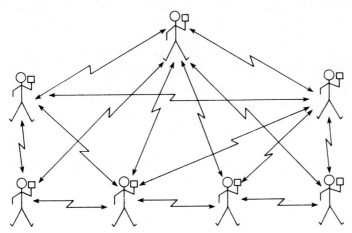

FIGURE 10-4
The Citizens Band (CB Radio) "Party Line."

users can reach an agreement that results in wide and apparently quite satisfactory use of the radio party line. A common CB language was developed which is still heard today in films and songs. But despite that often colorful language, the CB phenomenon soon wore itself out. The demand for licenses has fallen quite sharply, and there is little talk about additional bandwidth for the citizens band.

CB transmitters are usually amplitude modulated, as compared to the radio telephone transceivers described previously and in general use today which are frequency modulated. You will recall that frequency modulation tends to provide better signal to noise ratios, that is less noise on the channel. To conserve bandwidth, CB transceivers often operate on a single sideband. (Refer to Chapter 2 and Figure 2-12(c)). This not only conserves bandwidth—indeed, CBs can operate on one-third the bandwidth of units operating on both sidebands—it allows for higher power.

Unlike the radio telephone services discussed heretofore, numerous rules and regulations have been established by the Federal Communications Commission for what is, essentially, an open-to-everyone party line since licenses for CB use are very easily obtained. CB radio cannot be used for advertising, for soliciting the sale of goods or services, or for entertainment. (There are some who would argue that any operation of a CB radio is entertaining!) and users must watch their language!

Cellular Radio Systems

Would you agree that any service that: 1) requires you to wait almost five minutes to send a message, 2) gives you a busy signal two out of three calling tries, or even at best one out of every ten tries, and 3) requires that you use two different media in order to complete a conversation, as you do when you are paged, is a rather primitive telecommunications system, similar to the telephone of, say, seventy-five years ago? A major reason for these shortcomings is, certainly, the apparent shortage of spectrum that could be devoted to mobile services; so much of this valuable bandwidth has been taken up by radio and television as a result of the nation's desire for broadcast localism and the policies set forth to achieve this localism (as discussed in Chapter 9). However, we have seen how scarce and valuable bandwidth can be used more efficiently, primarily due to the computer-like functions introduced into communications by the availability of the semiconductor chip and the convergence of communications and computing. Earlier in this chapter we enumerated the techniques for making better use of bandwidth. Let's review them:

1 Frequency division multiplexing of the channel bandwidth.
2 Time division multiplexing of the channel bandwidth.
3 Allocating multiple channels as required (dynamic channel allocation).
4 Reusing the same frequencies.
5 "Real-time" management of channels.
6 Using higher frequencies.
7 Using digital signals on both analog and digital channels.

All but one of these techniques have been used for the provision of mobile communications services. Paging services frequency multiplex channels and send out digital pulses on the analog transmission, essentially time multiplexing the channel in order to send as many pages as possible per hour. Real-time frequency management is the job of the base station controller whether human or computer.

Radio dispatch and telephone services also use frequency division multiplexing and when sending coded "private" messages apply some limited forms of time division multiplexing. Increasingly, these radio telephone services dynamically allocate frequencies, in this way sharing or reusing the frequencies among different classes of users. And, of course, modern computer switching gear seeks to achieve a high degree of real-time bandwidth utilization, or channel loading as mobile radio-telephone enthusiasts say.

So far, we have not transmitted digital voice or data in order to achieve what is probably the most efficient utilization of bandwidth. Radio telephone using digital transmission is known as *packet radio;* we shall discuss this technology in Part 5 when we examine how we communicate with computers.

In our discussion of satellites we noted that frequency reuse is emerging as one of the more efficient means for getting the most out of a limited spectrum. We have seen that when dynamic channel allocation schemes are used for radio telephone services—the IMTS we discussed previously—waiting times were reduced and more radio telephone users could be accommodated. It is apparent that the key to making a limited band of frequencies serve a large number of users is to reuse the same frequencies over and over again.

Frequency reuse was applied long before the satellite or the radio telephone caught on to the idea. Recall the allocation of television channels discussed in

Chapter 9. We saw that the same television allocations were made to different cities as long as these cities were widely enough separated and the signal transmission patterns designed to avoid any interference. For example, consider the following allocation of VHF television stations to three "adjacent" metropolitan areas on the east coast (from Bowers, et al., 1978):

Metropolitan Region	Assigned TV Channels	
New York	2	4 5	7	9	11	13
Philadelphia		3		6		10	12
Washington DC			4 5	7	9		

Frequency reuse for cellular radio is accomplished by assigning one set of frequencies for radio communications in one geographic area and a different set of frequencies in the immediately adjacent areas. The first set of frequencies is reassigned to an area sufficiently removed from the first area where these frequencies are used to ensure that there is no interference. These areas are regular hexagons looking very much like cells in a beehive, hence the term cellular radio. The cells are usually a few miles in width, but their "radius" varies with the amount of mobile services in the area. The cells are generally grouped into blocks of seven cells although up to twelve cells are now being assigned to a block. The pattern of frequencies is repeated in the block, but within one block no frequency is allocated twice. A typical seven-cell block arrangement is shown in Figure 10-5.

This pattern of cells and blocks is repeated throughout the entire geographical area to be covered, usually the metropolitan area for which the cellular radio license has been awarded as, for example, the Chicago area (shown in Figure 10-6).

FIGURE 10-5
Cell Arrangement for Mobile Radio: A Seven-Cell Pattern Frequency.

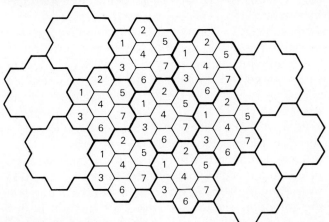

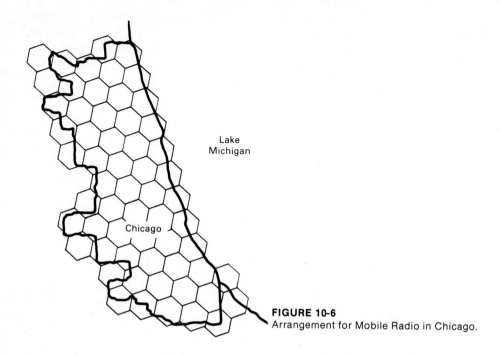

FIGURE 10-6
Arrangement for Mobile Radio in Chicago.

In each cell there is a low power base station. Each base station is interconnected through the main control or central station by landlines, either the twisted pair of the telephone or the coaxial cable of cable television systems. Microwave and even optical links also can be used to connect the cell base stations to the central control station.

When a mobile unit in a cell makes a call from any cell in the entire region served by the cellular radio service provider, the signal is carried by radio to the cell base station and relayed to the central station. The central station must then determine the cell of the vehicle being called and having done so routes the calling signal to the base station in that cell. The base station establishes the radio link with the receiving vehicle. This process is illustrated in Figure 10-7.

Since the cell base stations are usually connected to the central station by means of the local telephone operating company or to some other provider of telephone service that is required to interconnect to the local operating company, cellular mobile radio services can be fully integrated with fixed telephone services.

Just as the modern voice services we discussed in Chapter 8 would not be possible without microprocessors, computers, and memory systems, so cellular mobile radio services would be impossible to perform without these devices. This description of the calling and receiving process in a cellular radio system has exposed a host of control and signaling operations impossible to perform without the power of the computer and the common control concept first described in Chapter 2.

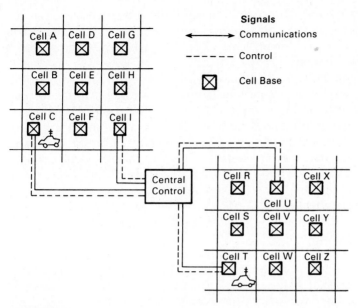

FIGURE 10-7
Making a Cellular Radio Call.

To begin with, when a caller signals the cell base station that a call is to be placed, the local base station must locate and assign an available two-way (*duplex*) channel and signal the caller to switch to those authorized transmission and reception frequencies. All of this must be done while the caller waits for a dial tone; the faster the cell base station can accomplish this, the better the service quality of the system.

Let's say the call is for vehicle or mobile M-2 in Figure 10-7. When M-2's number is dialed by mobile M-1, the central station sends out a paging signal to all the cell base stations which in turn send the signal to all of the vehicles in their respective cells, asking M-2 to identify itself. (When not in use, transceivers are automatically tuned to this paging channel.) M-2 recognizes its code and transmits a response signal.

The cell in which the vehicle is located must then be found, another important control function. Measuring the signal strength of M-2 is usually the way the vehicle is found, although detecting various other electrical characteristics such as phase shifts may be more accurate but more expensive.

Having located M-2, the cell base station assigns an available duplex channel from those authorized in that cell, and signals M-2 to tune to those channels. The conversation can now begin.

What happens if the vehicles talking to one another move across their respective cell boundaries? Not to worry. Memory systems at both the cell base station and the central station keep track of the caller's and receiver's identities. When the cell base stations detect the weakening of a vehicle's signal as it leaves the cell, the control equipment at the cell base and the central base prepare to assign new frequencies and do so at the appropriate time. The conversation goes on uninterrupted often with little or no apparent change in reception quality.

An important and early use of cellular radio communications was to offer travelers the opportunity to place telephone calls while on route by train between New York City and Washington, D.C. Business travelers, Washington lobbyists and those they lobby, and members of Congress were the targets for this cellular system. The fact that they could call ahead to confirm appointments and obtain the latest status reports before important meetings looked like a good business and a service that could attract passengers away from the air shuttles that travel between New York and Washington.

The 225-mile track between the cities is divided into nine mile zones of about equal length as shown in Figure 10-8. Each zone has its own fixed radio transmitter and receiver.

As the train moves from one zone to the next, calls in progress are automatically switched from one transmitter to the next; the speakers are unaware that the switch has even taken place.

The actual switching takes place at the central office in Philadelphia from where all of the transmitters and receivers in the nine zones (and the special ones in tunnels) are controlled. Switching takes place when the signal-to-noise

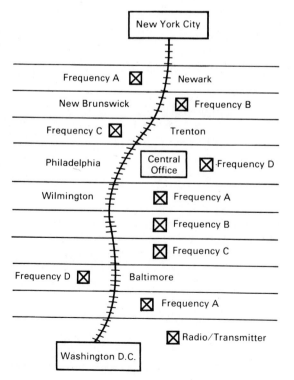

FIGURE 10-8
Metroliner Communications.

ratio falls below a certain level. This information is transmitted to the central office switch which searches for a transmitter with a better signal-to-noise ratio. When it finds one, it makes the transfer from one transmitter to the other; it does not have to know where the caller is on the route, only that a better signal-to-noise ratio transmitter must be found.

The major advantage of cellular systems is their capacity to handle greatly increased numbers of subscribers, while reducing the probability of getting a busy signal compared to what is usually found on the radio telephone system. We chose Chicago as the geographical area for illustration because until recently it had one of the more advanced multichannel mobile radio services and was saturated with fewer than 2000 subscribers. The cellular configuration illustrated in Figure 10-6 with four-mile cell radius now allows for more than sixty times that number of subscribers or about 120,000 mobile customers.

THE CONVERGENCE OF BROADCASTING AND TELEPHONY

It is most appropriate that we conclude our discussion of telecommunications for mobility with a description of cellular mobile radio services. Nowhere is the convergence of computing and telecommunications more evident than in the development of this technology. And nowhere is this convergence more evident than in mobile radio technology. Broadcasting has merged with telephony, a household-wired telephone now receives a call that begins its journey as a radio or broadcast signal and ends its journey on the twisted pair, and the called party is not likely to even be aware of this unless informed of it by the caller.

The technology of cellular mobile radio is most unique in that it calls upon all of the more sophisticated tools in the telecommunications and computing world today. From the world of modern voice services which we described in Chapter 8 we have borrowed all the technologies of common channel signaling—stored program control, store and forward, and those wonderful devices that can store and remember numbers, conversations, instructions, and whatever. From the world of broadcasting and especially the very "far out" world of satellites we have learned how to reuse frequencies, to dynamically manage them so as to make the best use of valuable and scarce bandwidth.

We can expect these development paths to continue as faster, smaller, and lower cost semiconductor switching and larger and lower cost semiconductor memories become available. Even if hardware limitations appear on the horizon, we have yet to exploit the programming software that can more efficiently manage the controls required for large-scale mobile telecommunications operations. The demand for mobile services continues to grow and this marriage of broadcasting and the telephone seems to have come along just in time to meet the demand.

THE ECONOMICS OF MOBILE COMMUNICATIONS

Economic planners, and especially those with a passion for telecommunications technology, have sought to quantify the economic benefits of investments in this technology. In the main, their searches have not borne much fruit; as an industry, telecommunications is of growing economic importance but attempts to measure the benefits derived from expenditures on the technology have been most difficult and, at best, only sophisticated guesswork.

Except in the case of mobile communications.

Here are some examples drawn from the many applications both in the United States and in other parts of the world:

> The Police Foundation reported that without two-way mobile radio, the number of patrol cars in the nation would have had to be doubled in order to provide the same level of services as the U.S. population grew and became more urban over the last twenty years.
>
> We have seen how the Soviets save time in the fields by equipping tractors and other agricultural machinery with two-way radios. The 20-25 percent savings thus realized by the Soviets has allowed for more training time to upgrade farm hands in new agricultural methods.
>
> The Polish government supported requests for additional mobile radio equipment with a report that during 1975 travel time from one place of work to another was reduced by up to 40 percent by equipping the vehicles of key workers with two-way radios.
>
> Trucking firms in the United States reported that the use of mobile dispatching systems enabled them to provide better routing of deliveries; productivity consequently increased by as much as 20 percent.
>
> Simply being able to monitor buses in the United States has reduced the number of buses required by 5 percent, according to statistics from several major cities. Monitoring also increased safety on public buses and reduced fatalities in accidents since time for assistance to arrive was reduced by at least 20 percent.

This should give some pretty good justification for investments in mobile telecommunications.

Numerous reports of the savings achieved from the use of mobile radio communications can be provided. The very fact that benefits can be quantified increases the demand for mobile communications services of all types. Because demand will continue to grow, costs for equipment will likely fall, thereby perhaps making mobile communications as available as the telephone is today. We have only just begun to explore the uses for these types of communications services; it is indeed fortunate that the technology has reached a level of sophistication where we can expect to satisfy these uses.

DICK TRACY WRIST WATCH RADIOS IN YOUR FUTURE

Science fiction writers and the comics have probably thought more about the future of mobile communications than have engineers and scientists in industrial laboratories. The ability to contact anyone anywhere in the world—indeed, in the universe—has been the classic ploy of many an adventurer in the clutches of an earth- or space-based enemy. Who has not read or heard of Dick Tracy calling for assistance on his wrist watch radio? Who is not thrilled when Buck Rogers or Superman calls on the trusty two-way radio strapped to his leg to seek help from allies in the friendly universe?

The technology to accomplish these feats is no longer fantasy; it is within reach today. All of the components are either in use or on the drawing boards and it is likely that within the next two decades we shall have Tracy's radio, if not on our wrists then certainly in our pockets.

The use of cellular radio in the conventional manner we have described has one major disadvantage: the mobile vehicle must be within line-of-site range of a base station antenna. For this reason, cellular radio coverage is likely to be limited to fairly large urban areas, areas with populations in excess of 200,000. Even the most optimistic estimates project that by the end of this century, at least 30 percent, if not 40 percent, of the nation's population will not be able to obtain mobile communications services. Clearly, this is an even greater possibility in nations throughout the world where greater numbers of people live in rural regions.

What if we could use a satellite as a base station for regions that are outside the line-of-site range of terrestrial base stations? What if we also could use the satellite as a central control for terrestrial base stations? We could then have an integrated terrestrial/land mobile satellite system, interconnecting all areas of the country with mobile services as illustrated in Figure 10-9.

FIGURE 10-9
Satellite-Mobile Communications. (From *IEEE Communications Magazine*, Nov., 1983, Vol. 21, No. 8, p. 13, with permission.)

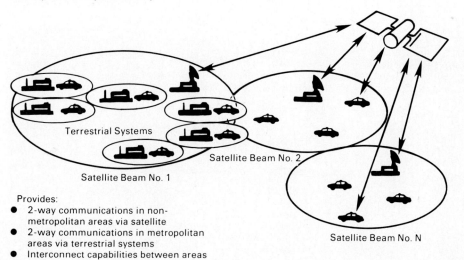

We learned in Chapter 4 that in the future, satellites will be capable of focusing powerful spot beams on very small footprints on the earth. What if we could cover the nation with a series of such spot or "pencil" beams, each beam acting either as a base station, as a central control for terrestrial mobile radio? The entire nation could be covered with cells corresponding to the pencil footprints as shown in Figure 10-10.

With the satellite also serving wire line telephone services and providing links to other nations and their wire line telephone and mobile radio telephone services, we can achieve the dream of an interconnected world—one that is linked by many networks, terrestrial fixed, terrestrial mobile, and satellite communications.

But we haven't yet realized the dreams of the science fiction and comic creators of Tracy and Rogers. If we continue to increase the number of computer and communications functions packaged into semiconductor chips as we have done over the past twenty years, there is every reason to expect that we shall have low cost, portable, and small cellular mobile radios before the end of this century. From there, it will be a very short leap to the pocket watch, if not wrist watch, cellular mobile radio transceiver with our own personal telephone number.

Science fiction and comic strip heroes are not the only beneficiaries of the advanced mobile communications on the horizon. Developing nations that cannot now make the very large investments necessary for a terrestrial wired telephone infrastructure could provide telecommunications throughout their entire countries without having to make the enormous investments in copper that the advanced nations have made. They could postpone these investments, perhaps to a time when lower-cost optical systems for their telephone infrastructure become available.

Mobile radio, the fastest-growing telecommunications industry in the world, provides another link in the networks that connect the world. Mobile tele-

FIGURE 10-10
Nationwide Satellite-Mobile Communications: Making Dick Tracy's Wrist Watch Radio a Reality. (From *IEEE Communications Magazine*, Vol. 21, No. 8, Nov. 1983, p. 14, with permission.)

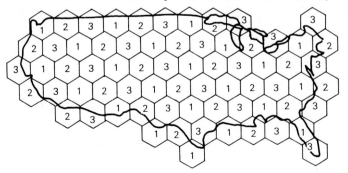

communications, whether as conventional dispatch, multichannel two-way radio, citizens band radio or cellular mobile radio communications, is linking nations in yet another global network, joining satellites, telephones, submarine cables, and broadcasting in what is clearly becoming a unified world of communications.

ADDITIONAL READINGS

The best way to keep up with this rapidly growing technology is to read newspapers and follow the advertisements for mobile telephone services. Articles and advertisements about mobile communications appear almost daily. However, the best single study of all aspects, societal, technical, and economic is:

Bowers, Raymond, et al., *Communications for a Mobile Society: An Assessment of New Technology* (Beverly Hills, CA: Sage, 1978).

PREPARING THE WAY FOR THE INFORMATION SOCIETY

So far in our study of modern telecommunications we have concentrated on technologies for the transport of information. Wires and electromagnetic waves are the carriers for the signals that transmit information, and we have explored the many forms these signals can take in order to perform this task. Now we turn our attention to the information itself and examine the many ways by which people look for and use information and how their information behavior creates demands for yet additional innovations in telecommunications technology.

We begin with the kinds of information seeking with which we are almost unaware—the casual way by which we discover what is going on around us and the very informal "searches" we undertake in order to make daily choices at home and often at work. Publishing without paper via teletext and videotex seeks to satisfy this human need. We also discuss the more purposeful search for the facts and figures we need to do our work using computers that communicate with other computers. Finally, we explore the emerging information marketplace on telecommunications networks, the network information services industry in the network marketplace.

CHAPTER 11

PUBLISHING WITHOUT PAPER: TELETEXT AND VIDEOTEX

When the history of twentieth century technology is written, it will be a story of fortuitous marriages and technological convergences resulting in industrial partnerships that in past years would have seemed quite difficult if not impossible. In our previous chapters we have seen how the computer and the telephone have come together to transform the POTS (plain old telephone services) into what has been called our most modern telecommunications system. We have seen the marriage of rockets and telecommunications for the launching of geosynchronous satellites that interconnect continents. Now we witness the joining together of broadcasting and computers, a partnership that may very well shape a new world of *electronic publishing*.

In this chapter we explore the technologies that make publishing without paper possible, teletext and videotex.

WHY PUBLISH WITHOUT PAPER?

How do you obtain information you need? When you want to know if today is a good day for tennis, a picnic, or just lounging around the house, where do you look? If you wish to purchase a personal computer, radio, or automobile, where do you go for information about the choices available? What kinds of questions do you ask of friends or relatives about restaurants, movies, or shows before you decide which ones you want to visit or see?

People seek information from a variety of sources; they hear and watch news and weather reports, see and hear advertisements on radio and television, read newspapers, magazines, and specialty publications, consult directories and timetables, and call up their friends and relatives. They sometimes visit libraries to browse through reference books. Information seeking is rather informal, often quite relaxed, and almost passive. We reserve active and formal information seeking for our business and professional lives where the searches are more intense, more specific, and detailed. In our work we are not merely interested in obtaining a general idea or an opinion, we want figures and facts which we cross- and double-check. In business we use a computer to assist not only to search for information but to help in the managing of it. Formal information seeking is what computers help us do; in the next chapter we shall discuss how communicating with computers helps us manage that information.

Despite the growth of home computing, however, there remains a big gap between the computer-based information systems used in business and the ways in which people informally obtain the information they believe they need. In business, the professions, and academia, computer-based information systems tend to be highly specialized, expensive to implement, and difficult to operate without special skills. Specialized information services have been developed to provide, for example, stock market quotations and other financial information for business customers and the numerous trade newsletters that monitor and report on products and services.

Can a system be devised that will provide information from hundreds of sources? That combines the speed and power of electronics with the familiarity of the written word? That uses friendly and familiar household tools such as the telephone and the television set? Can our homes and offices have access to an instant library, virtually unlimited in content yet tailored to each individual's needs? Can this library provide such services as airline schedules, stock market prices, up-to-date weather forecasts, and advertisements for automobiles and computers? Can we use this library to locate a particular restaurant in a particular part of town, to consult the menu and wine list and, if desired, to make a reservation from our home or office? Can we make the system as easy to use as our telephone and as easy to access as our television set, yet not as expensive as a business or home computer system?

If we were to do all of this, we would be *publishing without paper*. Teletext and videotex are technologies developed for just such a purpose. They are feasible because of the developments in modern telecommunications we have been discussing; they represent yet another fortuitous marriage of computing and telecommunicating.

FILLING UP TIME AND SPACE

In the previous chapter we examined the radio spectrum in considerable detail. We did so in order to understand the technological roots of the second era of broadcasting which has emerged over the past two decades, an era in which broadcast scarcity has been replaced with broadcast plenty. In particular we have seen that this development is having a profound impact on television, not only in the United States but throughout the entire world.

When new and improved broadcast technologies emerge, pressures are created on the managers of the radio spectrum to provide for new broadcast and narrowcast services. Rarely, if ever, are spectrum allocations denied, at least in the free market environment of the United States, and while it may take time for the political and economic actors to rationalize their positions, spectrum is almost always found and a new broadcast or narrowcast business is on the market.

Of all the services for which spectrum space has been allocated, television is the hungriest. While an AM radio station occupies but 10KHz of spectrum space and FM radio twenty times that much, color television requires 600 times as much spectrum as does AM radio. The television signal described in a previous chapter is a wideband signal made up of both the video and audio information transmitted. To conserve spectrum, television designers have already *filtered out* a portion of its lower sideband. Nevertheless, a moving image contains a great deal of information ("a picture is worth a thousand words"). To transmit this information so that we can capture that image to our satisfaction requires a bandwidth of 6MHz.

To ensure that the video received is the video transmitted, requires a number of control signals or *synchronizing pulses*. These pulses convey the information necessary for the scan lines on the receiver to work in *synch* with the scan lines on the transmitter. Now let us look very closely into the video signal of Figure 9-12.

The camera scans a scene once every 1/30 of a second. We want the receiver to do the same, so we send a *horizontal synch pulse* after every scan that creates the video signal—that is, the scene we are viewing. Three horizontal synch pulses are shown along with two "information" scanning lines in Figure 11-1(a). The actual horizontal synch pulse rides on top of a blanking pulse shown in Figure 11-1(b). This blanking pulse is strong enough to "black out" the electron stream retrace so that it will be invisible to the viewer. The horizontal synch pulses occupy a time period of 1/30 of a second, a valuable piece of time "real estate" in the radio spectrum.

FIGURE 11-1
Horizontal Synch Pulse: A Valuable Piece of Real Estate in the Radio Spectrum.

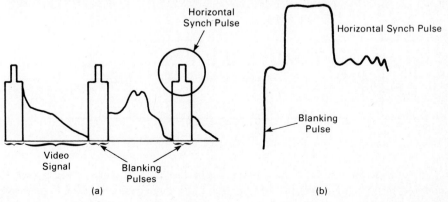

When the scanning for the first run across the scene is completed and the electron beam is in the lower right corner of the screen, the scanner must go back to the top and start over again. This movement of the scanning beam is called the *vertical retrace*. Since there are two runs across the scene and across the camera and receiver, there must be two vertical retraces. To make certain that the transmitter and receiver perform together again requires control information or vertical synch pulses. These are shown in Figure 11-2. The vertical synch pulses triggering the vertical retrace must occur after each 1/60 of a second since each of the two interlaced fields that make up one picture occur sixty times per second.

The vertical synch pulses and the equalizing pulses required to keep the scanning machinery under control make up the *vertical blanking interval*. The *vertical blanking interval* occupies 1330 microseconds, another valuable piece of time "real estate" on the radio spectrum.

These "properties" are portions of the control or synchronizing signals in the television signal; they carry information for control rather than for programming. There are other useful "properties" in the television signal that might be shared and thereby make more efficient use of this video bandwidth. For example, we could time or frequency multiplex the video signal but even if this did not require alterations of the many millions of receivers in the nation's homes, there could be some picture distortion, certainly enough to annoy viewers. Similarly we could time or frequency multiplex the audio signal in order to better share that portion of the spectrum. While this may not require massive changes in the hardware, it, too, might trouble listeners.

Over the years television engineers have sought ways to transmit a variety of special signals "piggy-backed" along with the television program. Examples of such signals include unobtrusive audio tones for alerting network affiliates to special announcements or for starting station videotaping equipment. Audio tones in ranges heard only by dogs are often used for transmitting control

FIGURE 11-2
The Vertical Retrace Interval with the Vertical Synch Pulses.

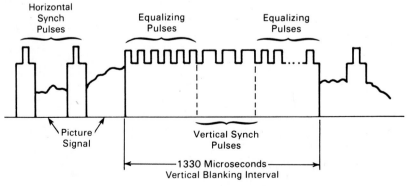

signals for AM stations. The vertical interval has been used for sending out test signals to monitor the quality of the video on the networks and for remotely controlling television broadcast stations. If you are alert you will sometimes notice strange flickerings in the upper right corner of your picture; they electronically identify the program or commercial during periods when sponsors are monitoring their investments. These signals, whether audio or video, must not interfere with your viewing or listening pleasure. The Federal Communications Commission and the broadcast industry have explored these signaling processes and have concluded that the vertical interval offers the best vehicle for the delivery of what are called ancillary signals, that is, signals that travel with the television signal but are not required for the audio or video transmission.

Teletext information is piggy-backed on the vertical blanking interval. Teletext is the one-way transmission of textual and graphic information to television sets along with the standard television broadcast signal.

HOW TELETEXT WORKS

Teletext was an idea of the British Broadcasting Corporation under the name of CEEFAX (See Facts) and by the British Independent Broadcasting Authority under the name of ORACLE (Optional Reception of Announcements by Coded Line Electronics). Both were introduced in about 1975 and started a wave of research in France, Canada, Japan, and other nations. (Teletext was a comparative latecomer in the United States although RCA tested a system to transmit text in analog form in the vertical blanking interval as early as 1960.) A major reason for embarking on these electronic publishing ventures was the desire to invigorate the British television and computer industries. By introducing new services, it was hoped a demand would develop for color television sets equipped with teletext capability thus accomplishing three goals at the same time; color television and a new service together could command a higher set license fee and raise more revenue for British broadcasting while at the same time, stimulate the development and production of microprocessors.

Teletext meets several of the major features we have suggested for a large instant library that can be tailored to individual needs; it can be low cost, friendly, and easy to use. It cannot, however, be used to make reservations since it is a one-way medium.

Teletext is possible because the picture and sound that a station sends over the air do not use all of the spectrum of the television broadcast signal. The unused portion of the television signal—that portion known as the vertical blanking interval (the black bar you see if the horizontal hold on your set goes on the blink)—is utilized for the transmission information that need not interfere with the regular TV picture. A computer converts printed material, including pictures, into digital signals that are transmitted along with the broadcast signal and are modulated or "piggy-backed" on the vertical blanking interval. The information is arranged as pages in a book, magazine, or newspaper. A viewer whose set is equipped with a special decoding device consults an index or table of contents and, using a keypad the size of a pocket calculator, selects a page. All of the pages are continuously being transmitted. The decoder selects the desired page out of the signal, provides new synch pulses to hold the signal in its local memory, and displays the information on the screen either superimposed over the picture or in place of the picture. This signal contains a page of information. The number of pages in the "newspaper" or "book" depends on how much of the vertical blanking interval or bandwidth the "publisher" is using and how quickly the publisher wishes the viewer to be able to access the pages.

You will recall from Chapter 2 in our discussion of information that the amount of information transmitted depends on the transmission time and the bandwidth used. The vertical interval is divided up into lines (time and bandwidth space) and the broadcaster can increase this time and bandwidth for transmission by using several lines. However, some of the lines are already being used by broadcasters for testing and signaling and if too many lines in the interval are being used, the coded pages of information "spill over" into the television picture creating interference.

In any teletext system there is the information provider and, of course, the information user, the viewer-reader at the television set. Providing information is not limited to the broadcaster. The broadcaster may act in a dual capacity; the broadcaster can be the major provider of information or may be a carrier of the information from many providers. Broadcasters, however, fear that they will be dubbed common carriers if they allow their vertical interval to be used by others. This creates one of the many contradictions confronting broadcasters who want to act like publishers. They want to continue their control over the content of their entertainment channel and, at the same time, have control over several information channels, a combination of roles that may fly in the face of the First Amendment.

Information is received or originated at the broadcaster's location; is formated for transmission and is inserted (modulated) on to the assigned lines in the vertical blanking interval. This formating can be performed at the information provider's site and telephoned to the broadcaster who will schedule the information for transmission.

Both teletext and videotex systems construct their databases to match the way we look for information, as pages in a directory, newspaper, magazine, timetable, or restaurant menu. This is considerably different from the manner in which computer-based information systems organize information. Business, professional, and research workers search for facts and figures rather than pages. Consequently, computer-based information systems organize information as numbers and words, by paragraphs or sections, and in charts and tables so that the users can manipulate them to suit their particular needs.

The competition among France, Great Britain, Canada, and now the United States has been over how information is formated for transmission. The graphic techniques and coding schemes determine the quality of the pictures produced.

Two basic graphical approaches are used: the British and French use an *alphamosaic* approach and the Canadians an *alphageometric* construction.

To create a teletext page using an alphamosaic scheme requires dividing the video frame into a number of blocks. Within each block the production terminal forms a matrix for the creation of the character to be generated. An 8 × 10 matrix is shown in Figure 11-3 which is usually provided for letters and numbers; a smaller matrix, only 3 × 2, is used for graphics. In the alphamosaic scheme texts and pictures are shaped by these matrices, hence teletext images tend to have the sharp edges characteristic of the illustration in Figure 11-4(a).

The Canadian alphageometric scheme uses picture description instructions to define basic shapes such as a point, line, arc of a circle, rectangle, or polygon. Further, the video frame is not divided into blocks; the shapes are created for each entire frame. Consequently, alphageometric teletext frames appear more rounded and somewhat more realistic as shown in Figure 11-4(b).

The scheme used depends, ultimately, on the size of the memory at the receiver. As with any broadcast service, the major cost of the system is borne by the consumer and if the cost of the receiving equipment is to be kept low, it is best not to require a large memory in the receiving equipment. Let's look into what the information user needs in order to read the electronic news.

Reading the Electronic Newspaper

The viewer switches his or her specially adapted set for teletext services and calls for the table of contents or index. Teletext transmissions are being transmitted continuously. If you were to "twiddle" with your horizontal hold and capture the black bar in the middle of your screen you would see white specks running along it; this is the coded teletext information.

The table of contents is delivered to the local storage. In an alphamosaic system, increased resolution and more rounded, realistic representations could

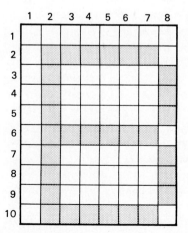

FIGURE 11-3
Creating the Letter "B" on an Alphamosaic Picture Frame.

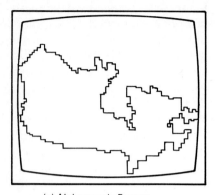

(a) Alphamosaic Format (b) Alphageometric Format

FIGURE 11-4
Competing Frame Production Formats.

be achieved by using more bits to code the alphanumerics and graphics in each block. (Recall that when we quantized our analog signal in Chapter 8 we found that the more bits used to represent the amplitude, the more accurate the digital reproduction of the analog signal.) However, this would require a larger memory in the local storage and a more expensive subscriber terminal. The alphageometric scheme selects the quality of resolution to deliver; the higher the resolution—i.e., the sharper and more realistic the representation—the more memory is required and the higher the cost of the terminal.

A major consideration in developing commercially feasible teletext systems is the design of this low-cost decoder. The decoder must be capable of storing at least a page of text (more would be better), sending instructions to the system, converting the data received by means of a character generator similar to that used by cable television systems to prepare data for television display, and providing the necessary synch pulses to control the stability of the picture. The decoder also may provide a variety of special signals to the viewer including signals that will display as many as seven colors. The more specialty features used to enhance the viewer's enjoyment of the visual experience, the more likely a viewer will take the trouble to "read" the television screen.

If teletext is to be the electronic newspaper of the future, it should possess some of the characteristics of a newspaper. One of these is a great variety of information available. It would be desirable to have very large libraries from which the subscribers can select for no two people read a newspaper for the same reasons. But the larger the library, the longer the reader must wait for the selected page to roll by. Remember, all of the pages are continuously and sequentially being transmitted. If, for example, a provider wished to offer 800 video or screen pages with about 120 screen words per page (this is equivalent to about a 50-page newspaper including advertisements) using two of the vertical blanking intervals which can achieve a transmission rate of about 13,000 bits per second, an access time or waiting time of almost 100 seconds between pages would be required. On the average the reader would have to wait more than a minute and a half to turn to the next page of his or her electronic newspaper.

Can you expect readers to wait a minute and a half before going on to the next page of a newspaper or magazine? Here is another example of the role of design tradeoffs. If you feel that no more than ten seconds would be acceptable, you could reduce the number of pages offered or use more lines. Indeed, if we use the entire television channel—that is, all of the lines in the video frame as we do in a cabled system, where an entire channel or 6MHz can be devoted to the newspaper—we reduce the access time to under ten seconds. That certainly sounds more reasonable.

Cabletext: Teletext on the Cable

Teletext over the cable has certain advantages that over-the-air teletext does not. Over-the-air we have only one broadcast channel with one television signal and one set of vertical blanking intervals. We have found that by using more lines in that interval we can increase the number of pages transmitted or increase the speed of transmission thereby reducing the access time. But remember, there is a danger of "spill over" into the broadcast signal if too many lines are used.

The cable, however, offers many channels and many more vertical blanking intervals. On the cable we can devote an entire channel with all of the lines in the video signal and all of the vertical blanking intervals to our electronic newspaper. Increasing the transmission speed from the mere 13KB/s to more than 6MB/s can make a significant difference as to whether or not teletext can be an electronic newspaper. But even more important is the fact that once wired, why not create interactive teletext or videotex?

Broadcast teletext is one-way. There are no transmissions between the information user and the information provider. Although it may seem that the user is accessing information in some data store, he or she is simply capturing and "holding" a small portion of the database in the local memory and interacts with that local memory. The information user does not interact with the data storage or database at the transmission site although the user can be prompted to search for specific information by means of a step-by-step tree-search procedure, as described in Figure 11-5.

The database is organized in a tree-structure. If you know the page you want to access you can do so directly. If you know that you want to travel by air to Paris, you can go directly to page six. The alternative is to begin with the master index and then select the pages you want by pushing the appropriate buttons on your keypad. Clearly, this is a slow process, but while going from page to page you can browse much as you do in an newspaper; you might even change your mind and end up going to Africa.

As the cost of local storage decreases and more pages can be stored in our decoder, pseudo-interactive teletext is possible. For example, consider playing games or utilizing programmed learning via teletext. The access time between

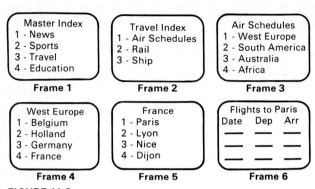

FIGURE 11-5
Planning Your Next Holiday with Videotex.

pages can be greatly decreased by "downloading" a lesson or a game in one fell swoop. The user calls for an entire series of pages and if the local storage has sufficient memory, it can store the entire lesson or game in that memory. In this way two-way interaction between the student or game player and the local storage is possible during the entire lesson or game.

There is, however, one significant drawback to broadcast teletext with a downloaded multipage search procedure for a game or for programmed learning. Providing a sequence of pages for a search procedure or learning program uses up a great many pages that could be more valuable to the broadcaster if used for one-way advertising. There is no way to bill the information user for the number of pages he or she uses; commercial teletext is therefore an advertiser-supported medium, as is broadcasting, and an advertiser will probably want to promote one-way messages of interest to the largest possible audience.

Why would someone engage in a rather long search process in order to obtain information about which he or she can do nothing? People who actively seek information as opposed to those who passively acquire it by browsing through a newspaper or asking a convenient friend do so because they have in mind a purpose for this information. One-way teletext does not provide a means to make the airline reservation or to reserve that table in the restaurant.

We learned in the previous chapter that cable television can be a two-way technology. If the information user is provided with the ability to access the entire database directly by way of the interactive capability the cable offers, an entirely different way of reading a newspaper or book can be achieved. Wired teletext made interactive by the ability of the user to communicate with the database is called videotex.

VIDEOTEX: THE WIRED NEWSPAPER

Teletext sits at the meeting point of broadcasting and publishing and, as we found, requires a switch from the video language of frames to the publishing language of pages. Videotex sits at the intersection of computing and publishing and to use it, we must be literate in the language of data acquisition, data processing, databases, and computer protocols as well as the language of pages, editing, and layouts. Videotex is often considered a computer-based information service suitable for non-computer users. When invented as Viewdata by the British Post Office it was to be a general information service designed to permit subscribers to use their friendly, familiar telephones and television receivers to access information much as a reference library would be accessed.

"An Interactive Information Service for the General Public," as Viewdata was proposed to the British Post Office in 1975 was an attractive investment for the telephone carrier. It offered an opportunity to increase the use of the telephone and create much needed operating revenue in a nation where personal telephone calling was still an event. It was, after all, an Englishman who, upon hearing about Bell's wonderful invention, remarked that the British did not need that contraption; there were plenty of messengers around. In addition to increasing revenue from the increased use of the telephone, a market would be created for computers, microprocessors, and software. Traditionally, British scientists and engineers had done outstanding research and development and had made several important computer developments, but no significant market had developed for their products. Viewdata, later to be dubbed Prestel (Press Telephone!) offered an opportunity to revitalize the lagging computer and telecommunications industries.

In principle, Videotex does not seek to perform the specialized functions expected of computer-accessed databases, the interactive services executives and professionals require to manage their business or perform research. (We shall examine these services in Chapter 12 as information systems for resources management.) Like teletext, videotex meets the requirements for a universal but specialized and easy to use information service; it delivers information as pages and uses the familiar, friendly telephone and television set.

But Videotex offers more than access to pages of information. Because it is interactive either via a telephone line or a two-way cable television system, videotex can perform transactions remotely, such as banking, shopping, and electronic message delivery. Videotex, like teletext, stores information in the form of pages; consequently, its transaction services are formal and structured. A shopper usually responds to a limited set of questions and has a limited number of prearranged responses; bargaining is therefore difficult. Early systems also provided a limited set of messages for electronic mail—birthday greetings, for example. However, today's systems allow for messages to be created by the senders.

Videotex differs from what we shall be discussing as network information services in Chapter 13 because it is designed for the general public and not the computer buff, for the user who has access only to a telephone and a suitably modified, low-cost television receiver. But as videotex increases the number of pages in its memory banks and decoders become ever more flexible and intelligent through the addition of typewriter keyboards, these differences may no longer be important.

Compare teletext and videotex in Table 11-1.

The principal differences between teletext and videotex are the latter's ability to offer two-way communication between information source and subscriber and the intelligence that this, in turn, offers to the subscriber. The intelligence is located at the carrier switching offices or the cable headend and can be shared among all subscribers, thereby reducing the cost of local intelligence.

Teletext and videotex both offer information in the form of pages and both deal with information as publishers rather than as designers of computerized databases. It should not be surprising, then, that the process by which information is prepared for delivery is the same for both videotex and teletext. And it is this process that further distinguishes these "mass information" media from computer-based data access systems.

As with teletext, pages of information are edited on a keyboard at one or more locations and delivered to a central computer. The design of the database and the associated software in the computer allows for accessing and rapid retrieval of specific pages as well as for billing the customers for those pages requested. The central computer is connected to the public-switched telephone network. A single computer, unless very large and with many access ports, might soon be overloaded and busy signals might deter persons seeking to purchase information. Consequently, a network of computers similar to the one shown in Figure 11-6 may be available to share information among them-

TABLE 11-1
COMPARING TELETEXT AND VIDEOTEX

Feature	Teletext	Videotex
Transmission	Broadcast	Telephone & Cable
Information Store	Limited	Encyclopedic
Retrieval	Yes	Yes
Information Sources	Broadcaster	Open to all
Response Time	10 seconds (av) for about 100 pages	22 seconds (av) for 100,000 pages
Mesages or Mail	No	Yes
Banking	No	Yes
Shopping	No	Yes
Problem Solving	No	Yes
Reservations	No	Yes
Usage Charge	No	Yes
Games	Yes, limited*	Yes
Programmed Learning	Yes, limited*	Yes

*By downloading into a local storage

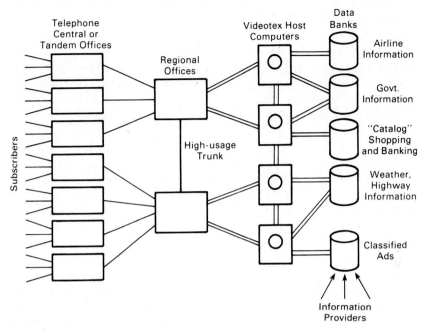

FIGURE 11-6
Telephone-based Videotex Network.

selves and to allow for regional access to that information. Interactive cable television systems also can access these computer networks.

Videotex information can be public or private. Firms may wish to provide data for their employees and can do so by requiring the information seeker to transmit a special code or password in order to retrieve this company-private information. Such systems are known as closed-user group services. Private line telephone access can perform the same function.

The subscriber facilities for videotex are quite similar to those required for teletext services; indeed, formats can be created so that the modified television receiver can accept both teletext and videotex information. Since transmission on the public-switched telephone network is analog, a modem is required to convert the signals into digital signals for storage or display. Videotex decoders have all the features of teletext decoders and in addition, they can communicate. The trend is toward a capability to receive data at a rate of 1200 bits per second and to send data at seventy-five bits per second. In addition, the videotex decoder contains special codes to enable the system to bill the user for the pages purchased. The decoder also must provide the video synchronizations to hold the captured video page on the television screen.

Videotex is among the more popular services offered to home computer owners, and indeed, much of the intelligence and communicating capability can be embedded in the personal or home computer. Protocols for transmitting data on the telephone carrier, which we discussed in Chapter 8, also are observed for videotex transmissions; personal computers can be programmed for a variety of transmission protocols.

Videotex pages are not cycled as are teletext pages. Users access the library page by page and pay for each page accessed. Page access time is determined primarily by how busy the computer is. But the number of pages stored in any single library certainly has some influence on the time it takes to find the information and this may create delays in getting into the computer.

The genius of teletext and videotex is in the creative use of the telecommunications and computer technologies. The multitalented semiconductor enables us to perform all of the classical functions of broadcasting more rapidly, with more efficient use of valuable spectrum, and at lower cost thereby opening up new uses for the empty spaces of broadcasting. Teletext and videotex have married these developments with the intelligent telephone. Our television set has become more than an entertainment device; it is now a tool of the new information age—an information appliance for the information household.

Teletext and videotex have unleashed a host of issues dealing with information policy. Information policy has not been a topic of serious discussion in our society; we have argued that our First Amendment, copyright and patent laws, antitrust laws, and the broadcast and common carrier regulations we have promulgated over the years constitute an information policy. In the final section of this chapter we ask if this is, indeed, so.

AN INFORMATION UTILITY FOR AN INFORMATION SOCIETY

Britain's announcement, in 1975, that it intended to embark on the development of general purpose public information systems using the telephone and broadcast systems was heard around the developed world. By 1981 the trade press carried discussions of Antiope and Teletel in France, Captains in Japan, Bildschirmtext in Germany, Telidon in Canada, and Swedish, Dutch, Finnish, Dane, Swiss, Austrian, Venezuelan, and Spanish versions of Prestel and CEEFAX. Why?

One common feature of these nations is that their telephone and broadcast industries are, if not entirely at least in part, nationally owned and operated. Broadcast revenues are obtained primarily through the levying of a tax or license fee on radio and television sets, although advertising as a revenue source is slowly creeping into these systems. The taxes and fees are voted upon by each nation's respective legislative bodies, an exceedingly unpopular task. Consequently, revenues often lag behind the budgetary requirements of broadcasters who are under public pressure to increase the amount of television offered and to improve its quality. Low-cost programs from the United States are popular with the public but are seen as diluting local cultures. However, without adequate funds local programs cannot be created. New uses for broadcast television can create new sources of revenue by requiring upgraded sets and associated hardware to be taxed at higher rates. Additionally, the demand for this new "high tech" hardware may stimulate local industry to enter the information era. And there are social benefits, too, that come with the equitable availability of information by means of the mass media.

In many of these nations, the telephone is perceived as an instrument for business rather than for households; telephone penetration ranges from 60 percent to 80 percent (except in Canada where it is well above 90 percent). Indeed, French parents, who have always perceived the telephone as an instrument for receiving news of serious disasters, are baffled by how quickly their children have adapted the telephone to their social needs rivaling those of the American teenager. Telephone infrastructures are highly capital intensive and require sufficient operating revenues to maintain, upgrade, and expand. Finding services that will increase the use of the telephone during nonbusiness hours of the day and encourage nonsubscribers to subscribe will develop much needed revenues. There are also social benefits to be derived from a public information system delivered by the public telephone system.

Teletext and videotex provide the impetus for the delivery of information services into the home. These emergent applications of telecommunications technologies have created the concept of an information marketplace, an idea quite radical and revolutionary in the minds of many people. Let's explore these radical notions for they establish the foundation of our final chapters on computer networks and the self-service society of the network marketplace.

If information is to be the currency of the information age then we must find ways of delivering it to the mass market. We have seen that the teletext and videotex technologies have ingeniously married the very latest technological developments to the familiar telephone and television set. What could be more conducive for developing a consumer market where there was none before? The range of information products offered and proposed is extensive as shown in Table 11-2.

It is left to you to determine which of these services are best suited to which of the alternatives, teletext or videotex. Clearly videotex offers the capability of providing direct two-way interactive services, but a creative page programmer can do wonders with the pseudo-interactivity available on teletext.

The emergence of electronic publishing has made social scientists more aware of the nature of information seeking. We know that social and economic factors influence how people seek information; middle class whites use the telephone more frequently than do blacks in the same economic category and senior citizens tend to depend upon television and radio for much of their information despite their self-reported high degree of dependence on the telephone. We have found that Latinos are more apt to turn to their community leaders for information even when bilingual information providers are available in city information offices and over the telephone.

We can classify information seeking as active or passive. Browsing through a newspaper, casually gathering information about department store sales, lazily watching the advertisements on television, and getting a general picture of

TABLE 11-2
INFORMATION PRODUCTS AVAILABLE VIA TELETEXT OR VIDEOTEX

Classification	Sample Information Products
General Information	Sports Results
	Weather
	Transportation News
	Community News
	Government Information
	Horoscopes
	Travel
	Restaurant & Food Guides
	Hobbies
	Telephone Directories
	Time Tables & Travel
	Recipes
Classified Ads	Real Estate
	Autos
	Jobs
	Schools and Training
	Services
Professional & Business	Facts and Figures
	Calculations
	Commodities
	Newsletters
	Reference Information
	Secretarial Services
	Accounting
	Technical Guides/Data
	Travel Reservations
	Restaurant Reservations
Messages	Electronic Mail
	Telex Interconnection
Shopping and Banking	Buying Information
	Money Transactions
	Selling Information
	Window Shopping
	Catalogs & Mail Order
Education	Home Study
	Homework Assistance

new automobile models can be classed as passive information seeking. When we go out of our way to make a telephone call to inquire about the price of an automobile or the availability of a product in the department store, or even call a friend to obtain additional information about an event we might wish to attend we are involved in active information seeking. The distinction becomes more important when we find that we must pay for the information we wish to obtain.

When we buy a newspaper for, say, thirty cents we are buying information. When we watch television or listen to radio and make decisions about products to purchase, we are buying information. When we make a telephone call to the travel agent we are paying for the information we receive in the cost of the telephone call. Yet most of us are unaware of the economic value of information and of information as a product. The idea that information is a product for which we must pay creates many problems for a democracy, where it is believed that information is the basis for preserving the system. Information as a product that may be priced out of the reach of some citizens troubles many, yet we have always paid for information, even for so-called "free TV," in the price of the goods we buy. Electronic publishing by videotex and teletext is a way of making information universally available at affordable costs in a world where information has become a commodity for which we must pay.

There will still be alternatives to electronic publishing; as we have seen, a new medium does not always replace the existing media. Films and radio did not disappear with the coming of television. Similarly we cannot expect the print media to disappear; each information and entertainment medium will serve the varied needs of society. Consider, for example, the weightings a sample of potential purchasers of electronic publishing gave to the several characteristics of four information/entertainment delivery methods shown in Table 11-3. Clearly, there is a place for all electronic modes for information seeking.

We alluded earlier to the collision between regulatory regimes that often accompanies a new technology. This collision is particularly noticeable in the United States where we have established clearer distinctions between the print and electronic media than have been established in countries where the

TABLE 11-3
HOW POTENTIAL USERS EVALUATE THE MEDIA

Feature	Print	Broadcasting	Teletext	Videotex
Portability	4	2	2	0
Selectivity[1]	3	1	3	4
Versatility[2]	4	3	2	4
Entertaining	4	4	3	2
Up-to-date	1	4	4	4
Easy to Distribute[3]	1	4	3	3

Scale: 0 very poor
1 poor
2 fair
3 good
4 very good

[1] The ease with which a searcher can retrieve specific information.
[2] The varieties of information that can be offered.
[3] Given that the subscriber hardware is available.

government plays a more significant role in the media. Over the years we have carefully defined the rights of broadcasters and newspapers and we had, until 1984, clearly established the position of the common carriers in the nation's communications infrastructure. Electronic publishing, however, is bridging these carefully crafted boundaries; print is available on the television screen, delivered by broadcasting and by the telephone carriers. The press is protected by the First Amendment, but broadcasters are not and must abide by a host of rules to ensure that the "scarce" spectrum is used in the public interest. Common carriers are prohibited from originating information that is carried over their wires, cables, and satellite beams. Yet storing this information in telephone company computers, and reorganizing and formatting it for distribution to videotex subscribers comes very close to information creation. Overseeing the telecommunications industries are the Department of Justice's antitrust laws making certain that industry concentration will not reduce the plurality of information providers necessary to our democratic marketplace of information and ideas.

The policy regimes and the rules they generate influence the electronic publishing industry and determine how it will develop. In nations where broadcasting and telephones are government-run monopolies, it is relatively easy to embark on a national program for electronic publishing, as have the Europeans and the Japanese. In the United States, however, the industrial actors come from different worlds and the rules of the game are often confusing and contradictory. Broadcasters, the newspapers and magazines, and the telephone carriers are uncertain as to where they can and cannot tread.

Electronic media allow for more knowledge, are easier to access, and can offer freer speech than we ever enjoyed before. As Ithiel de Sola Pool notes in *Technologies of Freedom*, "They fit the free practices of print." How we shall deal with this abundance is the issue before our policymakers, for, as Alexis de Tocqueville warned, democracy is often overzealous in regulating evils. Listening to Pool again, "... But as long as the First Amendment stands, backed by the courts which take it seriously, the loss of liberty is not foreordained. The commitment of American culture to pluralism and individual rights is reason for optimism."

ADDITIONAL READINGS

Teletext and videotex are relatively recent developments. Whether there is a market for the services they can deliver is an unanswered question. Nonetheless, the literature that has been devoted to the potential for electronic publishing is, to say the least, overwhelming. Much of the non-electronically published words today can best be summed up as promotional. However, there are several books that discuss these technologies expertly and fairly and leave the marketing hype to others.

Gross, Lynne Schafer, *Telecommunications: An Introduction to Radio, Television, and the Developing Media* (Dubuque, IA: William C. Brown Company, 1983). For a very quick overview of some of the latest experiments and demonstrations, see pp. 170–171.

Miller, Gary M., *Handbook of Electronic Communication* (Englewood Cliffs, NJ: Prentice-Hall, Inc., 1979). In particular, see pp. 270–276 for an excellent discussion of blanking pulses, vertical and horizontal synchronization, and other matters that are important to the understanding of how the teletext and videotex signals are squeezed into the television picture.

Neustadt, Richard M., *The Birth of Electronic Publishing* (White Plains, NY: Knowledge Industry Publications, 1982). A short preview of the potential legal conflicts arising from electronic print competing with traditional print.

de Sola Pool, Ithiel, *Technologies of Freedom* (Cambridge, MA: Harvard University Press, 1983). The last two chapters on electronic publishing and policies for freedom are especially good for understanding the dilemma created by the telecommunications and computer technologies and for an optimistic view of our future.

Siegel, Efrem, ed., *The Future of Videotex* (White Plains, NY: Knowledge Industries Publications, 1983). A realistic view of the future of electronic publishing.

Tydeman, John H., H. Lipinsky, R. Adler, M. Nyhan, and L. Zwimpfer, *Teletext and Videotex in the United States* (New York: McGraw-Hill, 1982). The most comprehensive view of teletext and videotex in the U.S.; this might be compared to the Winsbury book about the United Kingdom.

Winsbury, Rex, ed., *Viewdata in Action: A Comparative Study of Prestel* (London: McGraw-Hill (UK) Limited, 1981).

CHAPTER **12**

COMMUNICATING WITH COMPUTERS

We turn our attention now to the purposeful and more formal search for information. This is the world of information and resources management, time-sharing and teleprocessing, database access and electronic messaging, of worldwide airline reservation and inventory management, military defense networks, and strategic and tactical data management and control, the world of distributed data processing and data communications where computers communicate with computers.

We have traveled a long road toward understanding modern telecommunications and now find yet another way to communicate, *communicating with computers*. The computer has shaped modern telecommunications; without the logical processes that a computer can perform rapidly, repetitively, and accurately, and without its ability to store vast amounts of data and produce this data as the information we need to manage and control communications the communications revolution we are experiencing could not have occurred. The intelligence we attribute to telecommunications networks and terminals is due to the computer functions that have been deliberately designed into these systems and devices.

THE EXPLOSIVE INFORMATION EXPANSION

Telecommunications alone may very well have revolutionized our lives. Computers alone might well continue to make extraordinary changes in our world. But together they have a synergistic impact that makes for rapid change and makes possible the information revolution, or, as many would have it, the information explosion.

We see this revolution in the rapidity with which the capacity of telecommunications facilities has increased and continues to increase. Bandwidths have expanded far beyond the megahertz regions of the twisted pair and the coaxial cable into the gigahertz regions (thousand million hertz) of optical fibers and satellites and on to the frequencies in the infrared and light regions where engineers are testing lasers as carriers of information. An intercity telephone cable carries more than 10,000 voice conversations; some of the larger ones carry in the neighborhood of 100,000 voice conversations. Optical fiber telephone trunks and microwave transmissions in the gigahertz ranges can transmit a quarter of a million or more voice conversations. Couple this capacity with time division multiplex digital transmission techniques and there is no foreseeable limit to the transmission capacities of telecommunications networks. Consequently, we should not be surprised with transmission costs that are essentially free. The cost of communications will be in the information we are communicating, in the cost of its processing, and in the terminals we access.

As with all new technological breakthroughs the skeptics ask what we plan to do with all of this communicating capacity. Remember the British executives who were not too enthusiastic about the telephone and the network broadcast executives who wondered why they should pay for a twelve transponder satellite when three or perhaps six (two for each network) would do just as well?

It would be rash and foolish technological determinism to argue that since we have all of this communications capacity we *must* use it. This argument does not apply. Our complex and interdependent society requires information, "the essential resource for economic and social exchange," in the words of Daniel Bell.[1] We are in the midst of an extraordinary information expansion rather than an information explosion. The quantity of information being produced and made available is not likely to experience a sudden violent growth and then subside. On the contrary, history shows us that it will continue to expand and at increasing rates.

Only in recent history has codified knowledge become useful. Indeed, the very idea of codifying knowledge is a relatively recent phenomenon, dating, perhaps, from the eighteenth century. The application of this knowledge for useful purposes—the engineer's professional and social responsibility—transformed technology and society only in the past seventy years. As Cyril Stanley Smith, a renowned metallurgist observed, "in only a small part of history has industry been helped by science. The development of a suitable science began when chemists put into rational order facts that had been discovered long before by people who enjoyed empirical diverse experiment."[2] Daniel Bell argues that the formalization of judgments and experience, and their processing for transmission to others, "the intellectualization of experience and technology" led to the information society.[3]

Following the emergence of the printing press, the growth of information was extremely rapid; consider, for example, that in 1893 the rotary press, invented in 1850, produced 96,000 copies of the *New York World* per hour whereas only seventy years earlier, the newspaper was averaging 2500 copies per hour.

We may not know a great deal about the growth of general information, but we do know something about the growth of scientific information. Derek de Sola Price reviewed the growth of scientific information in his *Little Science, Big Science*. He wrote that the first two scientific journals appeared in the mid-seventeenth century, *Journal des Savants* in Paris and *Philosophical Transactions of the Royal Society* in London. By the middle of the next century there were ten scientific journals, by 1800 about 100, and by 1850 something in the neighborhood of 1000. Today it is well nigh impossible to keep up with the number of scientific journals; some say it is near the 50,000 mark, others argue that it is well over 100,000!

We have every reason to believe that this rapid expansion of scientific information is paralleled in industry; high-tech industries such as computers and biotechnology depend upon the rapid transfer of information from the research laboratories to the factory and office. Modern science and technology are becoming increasingly specialized in their pursuit of more detailed knowledge. But when there is integration of knowledge or transfer across disciplines, scientific breakthroughs occur. This is as true in industry and business as it is in science and this creates a demand for storing information so that is can be easily accessed and communicated.

Business demand for information is evidenced by the rapidity with which firms have installed computers. The growth of the computer industry is already legend; in a relatively few years, no more than forty, we have gone through four generations of computers and are now exploring a fifth and perhaps a sixth. From the massive mainframes that filled entire floors of buildings to the personal computers sprouting on employee desks and on factory floors, it is abundantly clear that information is the new currency of the business world.

Businesses need to create and maintain records for payrolls, for keeping track of employee benefits, and for making certain that financial transactions are carried out and that clearances and credit are properly allocated. They need libraries of information and databases to keep track of the characteristics of populations in order to research their markets better and more rapidly identify their customers. Firms now maintain a "corporate history" to facilitate decision making (after all, those who do not learn from history's mistakes are doomed to repeat them) and to satisfy government regulatory requirements. Information is required for scheduling factory operations, determining optimum product mixes, for analyzing and maintaining inventory, for scheduling transportation, and for making airline reservations.

The manufacture of aircraft, equipment for nuclear power generating plants, personal computers, and video games requires enormous amounts of information: words, numbers, and pictures. Greater product varieties to satisfy a multitude of human tastes as determined by finer and finer market analyses create the need for vast quantities of information that must flow through development, design, manufacture, sales, and inventory operations. Keeping track of variety requires information, everywhere. The many political promises to reduce the "size of government" have led to increased investments in the capital machinery necessary to perform the information activities required at the many levels of governance in our nation. Computers must gather and process the data that describe and define the nation; the economic planning of business and the social planning of the nation hinges on the information derived from the census and from the many other information producing activities of government.

Research laboratories and think tanks add their findings to the growing stock of information, a stock that information economists say is increasing geometrically.

We could go on, but it should be clear by now that we are experiencing an enormous increase in the amount of information produced. But if information is to be the new medium of exchange it must, like the old currencies of wheat, rice, oil, and money, be transported to where it can be used. Just as we developed highways for transporting goods and delivering services, so telecommunications networks are required to transport information products and services.

The 1960s saw an industrial expansion never before experienced in the United States, and indeed, throughout the industrialized world. Firms expanded and decentralized their facilities to take advantage of new population distribution, and states, cities, and nations offered attractive inducements to new industries in order to leap aboard what has become the high technology bandwagon of the 1980s. Because information needed to follow these industries wherever they might go, there was an accelerated demand for data communications. Businesses found it necessary to talk to other businesses; this increasing flow of information within and between business enterprises is what characterizes an information society.

Improvements in productivity in an information economy hinges on the quality of information and on how well it is used. Interconnection of the office's intelligent equipment and the factory's robots to the information required to make decisions and manufacture products in a timely manner is seen as the key to capturing the edge over the competition. Productivity in the office and factory is enhanced by the ability of computers to communicate.

Moving and manipulating the large quantities of information modern science and business require generate demands for ever more innovative telecommunications technology. We need to know how computers use telecommunications networks in order to understand the rapidly growing world of data communications. In this chapter we provide a preview of that new world.

THE COMPUTER FAMILY TREE

When computers communicate with one another they assume many roles; they can be relatively simple input/output terminals with varying degrees of intelligence or highly intelligent "hosts" performing complex multiapplication, multiprocessing functions. A comprehensive discussion of computers is not appropriate for a text on telecommunications; however, it is necessary to understand the computer's family tree in order to understand how and why computers communicate with one another.

Babbage could have built a general purpose mechanical computer in the nineteenth century that would have done much of what the first electronic computer, ENIAC, did in 1964. But he couldn't grind the gears accurately enough. ENIAC (Electronic Numerical Integrator and Calculator) was a very large system whose main job (and the one for which it was best suited) was the lengthy and repetitive calculation of ballistic tables for Army Ordinance. It was soon used for the equally lengthy and tedious tasks of calculating weather predictions, atomic energy data, and cosmic ray studies and generating random numbers and wind tunnel designs. ENIAC was a large, slow, difficult-to-program, general-purpose scientific computer. With the addition of stored program capability—a momentous development by that computer genius—John Von Neumann, the computer became more flexible, more general purpose, and somewhat easier to program. Many programs could be written in advance and stored in the computer. The computer could then give itself "instructions" for the job to be done.

For many years this has been the model for general purpose, centralized business computing. The machines were large, consisting of a *mainframe* which housed the primary components for storing information, processing programs, and calculating. They were surrounded by *peripherals* for printing output, storing data files, and inputting data via punched cards, tape readers, or keyboards. These mainframes required hordes of programmers, large numbers of staff to feed information into them, and a considerable maintenance team. They were very expensive and only the largest companies and the government could afford to purchase them; most were leased.

Mainframes, as they are still called, were rather impersonal devices; how can you be "friendly" with a giant that often consumes enough power to provide electricity for a good-sized town and requires several floors to house? These computers were operated in "batch" fashion—that is, you packaged your data on punched cards or on a magnetic data tape, walked it over to the rather frightening computer center, and humbly requested the intellectual giants who ran the place to do some computing for you. Several days later, sometimes weeks if we are to believe the horror stories people tell about the early computer centers, you returned and asked for your printouts. Heaven forbid if you were not happy with the results, even if the center was in error!

In time you were allowed to have a remote terminal—that is a terminal in an adjoining room—and you could insert the data yourself directly into the computer. If there weren't too many people sharing the giant you might even be able to do some of your work "on-line" or in real time. That is, you could see the results of the computation while you were "on the computer" and if necessary make corrections and modifications as you went along.

These mainframes are still with us and becoming increasingly important as terminals on computer-communication networks where they are known as "hosts."

Fortunately for us, the semiconductor chip and the photolithographic and chemical processes known as large scale integration (LSI) made the power of ENIAC available in smaller and considerably less expensive packages. At the same time it became clear to users that computers programmed to do a specific task or set of tasks would be easier to use and could be lower in cost than the massive mainframes that, supposedly, could do everything. Why should a firm wanting to do general business accounting—accounts payable, accounts receivable, billing, general ledger, and, perhaps, some inventory management and control—have to purchase a computer that could calculate missile trajectories and generate random numbers?

Small business computers or minicomputers soon came along. Compared to the mainframes, they were much lower in cost and very much smaller. Indeed, the input keyboard and the TV-like screen could be mounted on a desk with the storage files and power supplies in the same room or in a small closet. These minicomputers were interactive; you could talk to one via a typewriter-like keyboard and it talked back to you via the TV terminal in your office. No longer was a massive amount of paperwork required to prepare data for inputing and delivering the packages or batches to the computer center.

When you needed to access larger data banks or required more "computer power" than was available in your mini, you connected up to the host (mainframe computer). In time the telephone lines became longer and minicomputers as terminals were liberated from the mainframe room and building. As firms established decentralized facilities throughout the nation and the world, communicating with computers over telephone lines, by microwave links, and even by undersea cables and satellites became quite common and the data communications industry was born. Computers were communicating with computers and the telephone system had to learn how to deal with bits and bytes. We saw in Chapter 8 how it does so.

You know the rest of the story. Chip manufacture using very large scale integration (VLSI) manufacturing processes make possible the design of more and more functions on a chip—tens of thousands of computer functions packaged on quarter-by-quarter inch thin wafers of silicon. Each year, more and more computing elements get squeezed into a chip and computing power doubles. It was inevitable that the small business computer, our minicomputer requiring half a medium size office, would eventually be squeezed into a desktop computer, containing storage, power supply, memory, and everything the ENIAC had.

This new, personal computer is a wonderful stand-alone, general-purpose computer only if the "general purposes" you want to perform can be accomplished with its limited storage and memory capacities. These capacities are expanding with almost every new model coming on the market. However, as personal computers become more and more portable storage and memory must sometimes be sacrificed. The personal computer is a very intelligent terminal that can communicate with hosts, with business computers, and with other personal computers. Communicating between personal computers and with minis and hosts is making communicating with computers universal among business firms, on university campuses, and even in the home. In fact, these communicating personal computers are making the *network marketplace* a reality, as we shall see in the next chapter.

WHY COMPUTERS COMMUNICATE

A network of computers acting as terminals as well as processing centers is called a *teleprocessing* network. Data is transported between terminals and a data processing center for an increasing variety of reasons. In our brief history of computers we remarked about the low popularity of "batch processing," the practice of personally delivering your computing requirements to the computer center. One way of avoiding those usually arrogant computer center "geniuses" is to deliver your computing needs via a telephone line or even a satellite link. In this way, data can be entered and retrieved remotely on the teleprocessing network. Remote data processing for maintaining control of inventory, gathering data for payroll and keeping track of sales is an often overlooked but cost-effective means by which small and medium-sized organizations can access computer power not otherwise available with their limited budgets.

Many of today's airline reservation systems are data entry, collection, and retrieval teleprocessing systems. Travel agents enter requests for airline seats, auto rentals, hotel reservations, and special meals at remote terminals for transmission to the airlines' central or *host* computer; within a few seconds the transaction is completed and the good or bad news delivered to the agent and customer. Modern banking at automatic teller machines is also a form of data entry, collection, and retrieval teleprocessing.

Not too long ago your credit card status had to be checked on the telephone by a sales clerk, often necessitating long waits during heavy shopping periods. How often was the clerk forced to give up and take the risk that your credit was good? Today, card-reading terminals transmit the information on your card to a central host where credit, expiration date, and automobile license number are checked against a master file containing your credit history and heaven knows what else about you.

The telegraph operator that switched your message from one network to the other in the message-switched network we described in Chapter 5 and illustrated in Figure 5-7(b) has been replaced by computers that can store and forward your electronic mail. Mail stored in one computer is often forwarded to another computer for delivery to the addressee when requested.

Time-sharing in real time or in a "conversational mode" provides general and special problem-solving capabilities using shared high-valued programs and databases. Remote nodes can be relatively "dumb" terminals requiring that all of the processing be performed at the central computer, thereby putting a considerable data transport load on the communications links. With the availability of mini and personal computers on the network nodes, you can "borrow" central databases and programs for use at your terminal and "return" this valuable data when the task is completed. A time-sharing teleprocessing network with intelligence at the nodes is often called a *distributed data processing* network.

Personal computers are, increasingly, the processors of intelligence at these network nodes. With the increasing power of computers rapidly becoming available in smaller and smaller packages, why communicate with other computers? Why distribute computing?

The Great Debate: To Centralize or Decentralize

If there is any sort of heated debate in the data processing business it is over the question of whether it is better to centralize or decentralize computing. It is not uncommon to find firms shifting from one stance to the other as the cost of computers and communications changes. In a single organization there are often remnants of both types of operations, staff, management, and equipment. The politics of computing has resulted in many a fierce battle over whether to centralize computing and corporate power or to allow that power to be scattered. There are some who argue that it makes little difference, for however computing is distributed it does not change the power relationships in an organization. It is often forgotten that computers do not make decisions about computers, people do.

A centralized computer communications system has the network architecture shown in Figure 12-1. There is a single host computer, usually a large mainframe although in more recent systems the host might consist of several computers arranged as a multiprocessing facility. At each node there are input/output terminals with little, if any, processing ability. Note that the network shown in Figure 12-1 is a single level star network of the kind we discussed in Chapter 5.

FIGURE 12-1
A Centralized Computer-Communications Network.

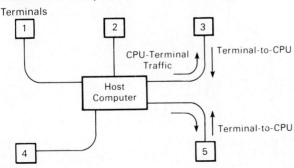

A decentralized computer communications system has the architecture shown in Figure 12-2. There are several hosts or multiprocessing facilities which may be mainframes or, in recent years, minicomputers. At each node there are input/output devices and multiprocessing facilities which also can act as input/output terminals when they access the processing services of other facilities. Figure 12-2 is a multistar network similar to the network shown in Figure 5-5.

While both the centralized and decentralized networks are similar in that they connect terminals and computers, there is a significant distinction between the two which becomes important when deciding whether or not to centralize or decentralize. Note that in Figure 12-1 only terminals are connected to the host computer or as network architects say, there is terminal-to-central processing unit (CPU) traffic. Because all of the traffic between a node requesting and receiving data processing services travels on a single terminal to the CPU channel, the transmission capacity of that channel must be large and, therefore, expensive.

In Figure 12-2, on the other hand, there is terminal-to-terminal traffic as well as computer-to-computer traffic and the likelihood that any single channel will be overloaded is remote. Hence, smaller channel capacity is required and the transmission costs per channel can be less. But more channels are required. Let's compare centralized and decentralized computing.

In a centralized system:

1 All applications are performed on one large host computer.

2 All the data is managed at the central location, consequently the communications protocols can be standard and relatively simple. Furthermore,

FIGURE 12-2
A Decentralized Computer-Communications Network.

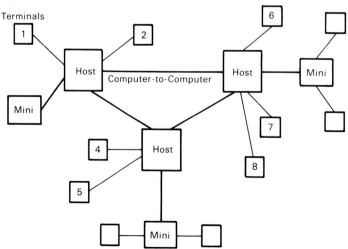

because the database can be centrally managed, its structure (how it is organized) can be standard and secure.

3 Since all the applications or uses are controlled from the central computer, new ones can be created relatively easily and quickly; no complex coordination with other data processing facilities is necessary.

4 If the processing system breaks down or "crashes," you know precisely where the fault lies—at the host—and recovery can be facilitated.

5 Certain economies of scale are possible—that is, with a large central host computer, greater investments in more sophisticated computing capability are likely to be made; the host has more dollars to invest than do smaller computer processing centers at many locations.

6 While no switching is required, the transmission paths must be broadband—that is, they must have a very large capacity with the attendant higher cost of such broadband transmission.

What about a decentralized system?

1 Applications are shared among many computers or processing centers; each may be expert at some specific class of functions.

2 If there is a failure at any location or even at several locations, the combination of functions performed at the remaining operating centers could continue to carry out the required tasks.

3 In a centralized facility there is complete reliance on the transmission channel from the remote site to the host. In a decentralized system, there are many channels to many processing sites; greater availability and less reliance is placed on transmission lines as well as on individual computers.

4 Decentralized facilities satisfy a need for local autonomy which may be of considerable importance in government as well as in corporate applications. As firms move around the world, this autonomy may be needed in order to satisfy host country requirements for doing business.

5 In a centralized facility processing tasks are shared with several locations; consequently, the bandwidth of the transmission lines between nodes can be less than the bandwidth required between any remote terminal and the central host. This reduces communicating costs; however, switching is required.

The battleground for the debate between centralization and decentralization is likely to be the telecommunications system and its costs. Data processing costs are decreasing rapidly as better and better semiconductor processing and memory chips are developed and produced. The cost of the processing complexes in Figures 12-1 and 12-2 are likely to continue to fall. The system cost will, therefore, depend upon the cost of communications as shown in Figure 12-3. As we noted in Chapter 2, the higher the computing speed, the more transmission bandwidth is required, and bandwidth costs money. However, we have found that with time division multiplexing techniques, more efficient use of the channel is possible. We also can expect that broader bandwidth channel capacities will be available as we move from coaxial cables to fiber optics and laser-driven communications. Remember the technological

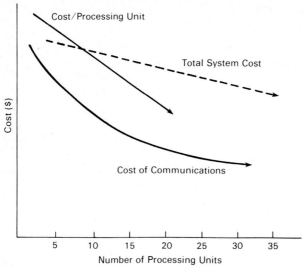

FIGURE 12-3
Cost Trends Favor Distributed Processing.

breakthroughs we discussed in Chapter 9 that will open up the lightwave and even infrared bands in the electromagnetic spectrum making available even higher frequencies and broader bandwidth. As a result, system costs per processing unit are likely to fall over the next several years and the trend will continue to be toward distributed data processing.

DISTRIBUTED DATA PROCESSING

A network of computers and input/output devices where either:

   1 The same function can be executed at more than one point or node in the network, or
   2 The function cannot be efficiently executed in a single node and portions of the process must be executed or shared cooperatively among nodes,

is a *distributed data processing system*.

At every node on a distributed data processing network terminals have the ability to call for information from another node and to allocate some function which they may not be able to perform to another node. Each node has decision logic processing capability—the ability to make choices, to pass information along to other nodes or to the hosts, and to perform data processing itself. Some nodes on a distributed data processing network are simple input/output terminals with limited intelligence.

Interaction with distributed data processing systems takes place every time we make a travel reservation or bank from a remote automatic teller machine. Information systems for making management decisions are distributed data

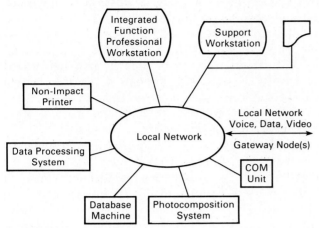

FIGURE 12-4
A Local Office System Network Is a Form of Local Area Network (LAN).

processing systems. They may operate worldwide, as do those of multinational firms, but they also can operate within limited geographical areas in an office, a building, or an industrial complex. Today, these networks are referred to as *local area networks;* they distribute data to nodes where processing may take place or to terminals where executives can play the "what if" simulations decision making requires. A local area network for an office is illustrated in Figure 12-4.

Distributed data processing systems can be classified along the following lines:

• Special purpose networks, usually having a single application and used by a single organization. A firm with several sales offices scattered throughout a city may want to interconnect these offices to keep track of sales and cash flow, and to enable better control of inventory. Major national merchandisers collect the "receipts" at the end of every day in a central location in order to control cash outlays and better manage inventories which, if allowed to get out of control, could create considerable financial problems.

• Early distributed data processing (DDP) networks were quite similar to the switched Telex or teletype systems in operation in many places throughout the world. The central processor or host receives information from the terminals and schedules this information for processing in the computational facilities of the host computer. When the processing is completed, the host sends this information on to another terminal or back to the originating terminal. One way of looking at a DDP system is as a message-switched network (see Figure 5-7(b)) where the messages are requests for data processing. Some airline reservation systems are single purpose, single application systems by which travel agents at remote terminals signal reservation requests to a central host computer and wait for confirmation of those requests.

- Increasingly, however, these special purpose networks are being transformed to multipurpose, multiapplication networks which perform several tasks required by the parent company. Firms add to their cash monitoring functions the task of scheduling deliveries out of their main warehouses, placing orders to replenish inventories, and maintaining the personnel records of thousands of employees. Airline reservation systems which provide seat reservations and related travel services such as auto rental, hotel reservations, and travel packages to their own reservationists, travel agents, and other airlines are examples of more general purpose network systems.
- The final step in the evolution of computer-communication networks concerns those computers that share resources among many diverse users for many diverse uses. Perhaps the most famous of these is the Advanced Research Projects Agency (ARPA) network interconnecting highly complex and often incompatible computer processing facilities throughout the nation. The ARPA network is a supernetwork, connecting terminals, computers, and data processing centers equally. It has become the model for the many commercial computer-communications networks now circling the globe, including the international travel reservation system, SITA (Societe Internationale de Telecommunications Aeronautiques) and the worldwide banking network, SWIFT (Society for Worldwide Interbank Financial Transactions).

Information Systems for Resources Management

Many of today's management information systems are distributed data processing networks. Our repeated references to airline reservation systems are not purely casual. We cite them because they represent some of today's most sophisticated yet universally available distributed data processing systems for managing resources.

Consider what it takes to run an international airline system. The travel industry is traditionally service based and labor intensive. Its quality depends upon how fast, accurate, and accessible these services are. It is, to a large extent, an information business—selling information about schedules, seats, the aura of romantic destinations, and the promise of adventure. An airline must coordinate a multitude of transactions, people, and activities including passengers, crew, cargo, equipment, parts, maintenance and service personnel, food, fuel, lodging facilities, financial information, systems planning, and scheduling. The system also keeps track of fares, taxes, and special rate considerations and performs credit checks on ticket purchasers using credit cards.

The system has a high degree of accuracy, and is capable of verifying, tracing, and documenting reservations and other passenger services; it even recalls transactions so that a transaction history can be traced if desired. In addition, many airline reservations systems perform accounting and personnel functions for their travel agents and others who wish to purchase these services. Some analysts have estimated that as much as 20 percent of an airline's operating costs is devoted to the management of these information activities.

The system's terminal and processing nodes are located throughout the world. Consequently, a wide variety of terminals and processing equipment communicates on an equally diverse number of telecommunications facilities. Voice-grade telephone, high-speed data transmission systems, low-speed telegraph networks, satellites, high frequency radio, and mobile radio communications are often used to interconnect a world-wide system. Terminal characteristics vary from those of computer-to-computer communications which could be in the megabaud range, to those of travel agents transmitting at 1200 baud, to the telegraph systems that transmit at only five words per minute. Interactive real-time conversations between agent and computer with response times of no more than three seconds are required in the major cities while remote sites served by telegraph usually receive responses in minutes.

An airline reservations system is, clearly, a multifunction, multiapplication distributed data processing system. Telecommunications facilities for serving this wide range of communicating needs require the judicious use of protocols and the mixed use of both message and packet switching. Often complex distributed data processing networks with wide ranges of speed and information presentation requirements are designed as two networks, a high-speed fast response system and a low-speed medium response system. Interconnection requires a store and forward messaging facility utilizing paper tape readers and human telegraphers. So you see, modern telecommunications has not yet outgrown the telegrapher with a green eyeshade or the punched paper tape.

TELECOMMUNICATIONS FOR COMPUTER-COMMUNICATIONS SYSTEMS

In Chapter 8 we discussed how to talk to a computer by telephone. We found that plain old telephone services, our friendly POTS, is changing its spots and becoming the primary network not only for voice-to-voice communications or even people-to-computer communications but also for computer-to-computer communications. We have not yet arrived at the ultimate integrated services digital network described in Chapter 8 where all messages, whether from a person or a computer, are identical bits. Most computer-communications systems use the public-switched network or private lines constructed for the system by the public carriers. For some years to come, these networks will be both analog and digital, as we found in our discussions about the intelligent telephone.

The primary objective of the public-switched telephone network is to enable subscribers to communicate with one another. Telephone users are satisfied with a fixed channel capacity between 300 and 3400 hertz. They can usually tolerate noise and still make themselves understood, always participate in two-way conversations except when storing or forwarding a message, talk or listen continuously until the conversation is completed, and are usually patient even if they must wait several seconds or even a minute and more to make the connection. The transmission rate is constant for a telephone conversation for the entire time of the connection. People talk to people over similar terminals. Reaction and response times, and the length of the conversations do not vary much. There is enough similarity and homogeneity to predict the likelihood of busy signals, blocking probabilities, and other measures of telephone service quality discussed in our chapters on the telephone.

Teleprocessing networks, on the other hand, are not homogeneous. Teleprocessing networks provide communications between often widely geographically dispersed terminals, databases, and computer processing centers operating at varying speeds or bit rates and on different protocols, and have vastly different message characteristics. Some terminals only transmit, others only receive, and some do both. Data must be delivered error-free; excessive channel noise will usually cause the connection to break down. Data is frequently transmitted in bursts, especially in a man-computer dialogue and when time-shared computers are "talking" to one another. Users expect to receive their information in different forms; some wish to receive printed output, others are satisfied with CRT displays, and some wish to edit their input "on-line"—that is, to compose their messages in real time even if the other party is not "on-line." Teleprocessing networks are distributed intelligence networks and the intelligence may be vastly different from terminal to terminal and from processing center to processing center.

Teleprocessing networks are required to provide a wide range of channel capacities or bit rates ranging from as low as ten bits per second for the telemetering used in monitoring and control, up to 1.5Mb/s for high-speed computer-to-computer data transport. If offices wish to interchange video, the network must be able to transmit at 96Mb/s. The range of transmission speeds a data communications network may be required to deliver is shown in Table 12-1.

Controlling Access to the Telecommunications Paths

In teleprocessing networks involving multiple terminals, hosts, and lines, efficient system operation requires unusually disciplined protocols to ensure access to and utilization of the networks. The effective utilization of multiple network topologies and hierarchies and time division multiplexing techniques are only the first steps in the design of an efficient data distribution or teleprocessing network. Several categories of control have been developed to meet the user needs for response time, accuracy, and reliability.

The simplest type of line control is essentially no control at all or what is known as *contention*. Terminals compete for access to the line, a process very much like sharing the old telephone party line. If a terminal requests access to an unoccupied line, the message is sent. But if the line is busy, the terminal waits until the line becomes free. The communication control program in the main computer host will often build up a waiting line for access to the network and allow access on a first-come, first-serve basis. Sometimes the control computer will set up a procedure for access according to a prescribed sequence. But contention protocols are not adequate for data distribution networks that

TABLE 12-1
TELECOMMUNICATIONS BANDWIDTH REQUIREMENTS FOR DISTRIBUTED DATA PROCESSING

Service	Data Rate
Telemetering for control and monitoring	10b/s-1000 b/s
Energy management	100b/s-1000 b/s
Interactive services	
(Banking & other transactions)	100b/s-64Kb/s
Electronic mail	4.8Kb/s-64Kb/s
Mass data transfer	4.8Kb/s-44Mb/s
Facsimile	4.8Kb/s-1.544Mb/s
Slow-scan video	56Kb/s-64Kb/s
Full-motion video	96Mb/s
High definition television	200Mb/s and up

are heavily loaded, especially if there are many nodes and many lines on the network. It is not at all unusual for an individual terminal to tie up a line for very long periods of time.

Polling is another form of line control frequently used on multipoint lines. A central host will selectively ask a terminal if that terminal has a message to send. If a terminal has nothing to send, it "says" so and the questioner moves on to the next terminal. If the terminal responds that it has a message to send, the central host does not move on until that message is sent. Each terminal has a unique address and responds to polling questions only when its code is sent and received.

While contention protocols impose the least amount of procedural requirements, especially on the software necessary to control line access, this form of line control can be highly inefficient. Polling requires more "overhead," as data communicators call the necessary software (and hardware) protocol techniques, but is more responsive to today's complex data communications needs.

Data communications has emerged as an engineering discipline on its own. While it is a proper subset of telecommunications engineering it is, clearly, beyond the scope of this book. We have, however, described the basic tools that are used in designing data communications networks: time division multiplexing, packet switching and transmission, digital transmission and digital switching, network architectures and distributed network intelligence, and the ever-present and most important task of setting protocols so that what is transmitted is intelligently received.

NETWORKING IS WHAT MODERN TELECOMMUNICATIONS IS ALL ABOUT

Computers communicating with computers have altered the shape of telecommunications throughout the industrialized world. To some considerable extent, the reshaping of the nation's telephone carriers has been the result of the demand by firms to communicate with their computers. This phenomenon is not limited solely to developed nations. In every nation where the information society has touched base—where planners believe that the road to prosperity through modernization must be paved with chips and computers—one hears the almost standard critique of the telephone carriers. "Private lines for data communications are not forthcoming and the public-switched network is too slow." "Line switching is inadequate for data transfer; we need packet switching or better quality message switching networks." And so on. The pressures on telephone carriers throughout the world for more rapid change have, to a great extent, arisen from the demand for improved telecommunications facilities for computer-to-computer communications.

REFERENCES

1 Bell, Daniel, "The Social Framework of the Information Society," in Michael L. Dertouzas and Joel Moses, eds, *The Computer Age: A Twenty-Year View*. (Cambridge, MA: MIT Press, 1980).
2 Smith, Cyril Stanley, "Metallurgy as a Human Experience," in *Metallurgical Transactions A*, 64, no. 4 (April 1975): 604.
3 Bell, Daniel, op. cit.

ADDITIONAL READINGS

In a surprisingly short time, computers have become a "necessity" for both business and government. Because of their rapid proliferation throughout so many different businesses and levels of government, scholars and journalists have been interested in the consequences of computer use. Hence, some of the most interesting and relevant readings concern the applications of computers and the social consequences of these applications.

Danziger, James N., W. Dutton, R. Kling, and K. Kraemer, *Computers and Politics* (New York: Columbia University Press, 1982). This study inquires into what happens in government when computers are introduced.
Dertouzes, Michael L., and Joel Moses, eds., *The Computer Age: A Twenty-Year View* (Cambridge MA: MIT Press, 1980). Probably the best overall, readable summary about "the computer age."
Doll, Dixon R., *Data Communications: Facilities, Networks, and System Design* (New York: John Wiley & Sons, 1978). See Chapter 8 for a more detailed discussion of "line control procedures, network protocols, and control software."
Kraemer, Kenneth L., W. Dutton, and A. Northrup, *The Management of Information Systems* (New York: Columbia University Press, 1981).
Meadow, Charles T., and Albert Tedesco, *Telecommunications for Management* (New York: McGraw-Hill, 1985). Chapter 17 describes a corporate management information system.
de Sola Pool, Ithiel, Hiroshi Inose, Nozomu Takasaki, and Roger Hurwitz, *Communications Flows: A Census in the United States and Japan* (Tokyo: University of Tokyo Press, 1984). More proof that we are in an information explosion/expansion.
de Solla Price, Derek, *Little Science, Big Science* (New York: Columbia University Press, 1963).

CHAPTER 13

NETWORK INFORMATION SERVICES: A NEW INDUSTRY

THE LITTLE HOUSE ON THE NETWORK

When is a house more than a home?

Today's house is a node on seven networks, six electronic and one concrete. Just about every house has a telephone and few if any households are not within reach of a television and radio broadcast signal. Those not directly "connected" to a broadcaster are reached by translators which retransmit television signals across the last mile. Almost two out of three households are passed by cable and one third of all television households is connected to the cable. No home in the United States, including those in Hawaii and Alaska, will be out of sight of one or more direct satellite broadcast signals by the mid-1990s. Households may even wish to be on the cellular radio network or several of the microwave options that "by-pass" the local telephone loop.

Yet another electrical network interconnects households, one that is so ubiquitous we often forget that it exists. The electrical power network has been used to deliver control signals for monitoring and controlling power usage by means of a *ripple current* that "rides" on top of the electrical power. In a slow pinch this network can even be used to deliver mail.

Finally we must not forget the highways and streets along which mail, newspapers, and people travel to deliver goods, services, and messages. Since, as urban planners tell us, we must learn to live in more densely packed housing we can expect this "traditional" mode of nonmediated communications to be used more rather than less. We might wonder about the need for a telephone in Bombay where word of mouth communications is often more rapid and reliable than the telephone network.

Count them! Count the multiplicity of networks along which people can communicate:

1 Telephone.

2 Cable television—54 channels and more, as many channels as desired and for which you are willing to pay, both via additional cables or optical fiber transmission paths.

3 Broadcast—radio and television; multiplexed subcarrier radio channels, low-power UHF television, and VHF "drop-ins" that more efficient use of the television spectrum can make possible, and multipoint distribution channels that increase the number of broadcast signals into a home.

4 Mobile radio via cellular radio and the microwave telephone "by-pass" modes become increasingly attractive for consumers as prices fall and especially as the cost of local loop telephone communications rises.

5 Direct satellite broadcasting, perhaps as many as six channels available to a household.

6 Ripple current along the electric power provides "electronic" mail in a pinch.

Not all of these pathways are interactive; however, mixed asynchronous use can provide interactivity. For example, it has been suggested that in the off-hours—the dead time of early morning, for example—broadcast channels can be used for the delivery of information to households and businesses authorized to receive these signals. Consider, for example, an accounting service by which a subscriber sends daily transaction information "upstream" to the broadcast station via telephone or cable where it is processed and during the "Off-hours" transmitted at very high speeds to the subscriber's personal computer. In a way this is similar to the downloading of games and programs teletext providers offer to simulate interactive transactions on that one-way service.

It is very likely that many of today's "one-way" technologies will learn clever tricks to become interactive. Relatively small and low-cost satellite systems for two-way data communications—the thin-route satellite networks we discussed in Chapter 5—are now being used for interconnecting business and personal computers both in the United States and in developing countries.

With so many opportunities to reach into households and the growing intelligence of transmission channels—intelligence made possible by computerized technologies for telecommunications—it is not surprising that merchandisers, information and financial services providers, and manufacturers envision a new marketplace. The network marketplace not only offers new ways of doing business but, perhaps more important, the means for creating new businesses and new educational, cultural, and social experiences.

The preceding chapters in Part 5 have discussed the technologies that underpin this emerging network marketplace. In this chapter we discuss how these technologies are creating a new industry.

THE ELEMENTS OF THE NETWORK INFORMATION SERVICES INDUSTRY

While sociologists and philosophers debate whether we are, indeed, entering or in the midst of an information revolution and whether we have or will become an information society, a network information services industry is evolving. Technologists and entrepreneurs do not wait for philosophers, but leave it to these philosophers to name the worlds they are creating. Before examining the components of this industry, let us briefly review what we said about the network marketplace and the information society in Chapter 1.

The marriage of telecommunications and computers, we wrote, permits users to interact directly with one or more computers, with distributed information systems, which may be in banks, shops, airline terminals, libraries, or schools. Manufacturers in foreign countries can operate facilities programmed by computers in the United States so that the latest models can be on the production floor within minutes of the completion of engineering testing. And should Cold War politics call for cutting off manufacturing on foreign shores, the flow of technological information can be cut off by the "flick of a switch." Information about money, as valuable and as trusted as money itself, travels across international borders, every transaction increasing the value of the money as exchange rates vary. Transborder data flows are the new products of international trade and no one is quite sure what this will mean to the international distribution of labor nor to creating a more equitable worldwide economic distribution of wealth.

The goods and services this network industry will deliver are only now becoming clear. Some services and products now becoming available are shown in Table 13-1.

The network information services industry is in the business of providing computer/communications-mediated information services. In its simplest form, a consumer or purchaser of information (and this may include the performance of a transaction that exchanges information and money) interacts on-line with one or more computers and associated information files from a remote terminal or computer and receives information of value in a rapid, economical, and convenient form. The remote "batch" computer operations we described in the previous chapter are the means by which the consumer purchases "raw computing power" for delivery to a facility where this raw material is used to create higher-valued information products for sale.

Financial planners purchase payroll packages, budgeting, planning, inventory control, market forecasting models, and data. Advanced scientific information products, such as statistical routines, analytical models for examining seismic and geological data for oil exploration, and bibliographic search procedures for oceanographers, are offered for sale. Physicians can "dial up" a cardiac consultant or hematological differential diagnosis model. The number of available information "raw materials" is growing rapidly.

Of course the number of network services is also growing as rapidly as human behavior adapts to this way of doing business and of carrying out daily household activities. Banking on the network is highly efficient; bankers say that the cost of a network transaction is almost one-sixth of its over-the-counter cost. You can expect all sorts of incentives to wean you away from the teller for routine transactions. Window-shopping via computer, a sort of electronic catalog on the television receiver, is waiting in the wings for some appraisal of

TABLE 13-1
SOME POSSIBLE NETWORK INFORMATION INDUSTRY PRODUCTS AND SERVICES

Business Applications	Medical Applications
Remote computing & analysis	Remote diagnosis
Mass mail selective advertising	Emergency medical information
Management systems	Medical management systems
Interfirm transactions	Patient information systems
Analysis and simulation	Remote computing & analysis
Personnel database maintenance and inquiry	Access to public health databases
Scientific & engineering analysis	Computer & teleconferencing
Remote graphics	
Electronic mail	
Cashless transactions	
Banking	
Legal services	

Government Applications	Consumer Applications
Remote Computing & analysis	On-line catalog shopping
Vocational counseling & placement	Working-at-home
On-line polling	Consumer advisory services
Remote court rooms	Utility meter reading
Vehicle scheduling & control	Hotel, travel, entertainment ticketing
Electronic city directory	Electronic yellow pages
Management systems	Electronic publishing
Traffic control	Dedicated, tailored newspapers
Pollution monitoring & control	Home protection services
	Personal data management
	Personal tax preparation
	Games
	Banking & bill paying
	Movies, concerts, plays, & opera

Educational Applications

Remote library access
Adult courses
Computer-aided instruction
Preschool instruction
Drill & practice & homework
Educational management
Gaming and simulation

customer's willingness to purchase on the network. If the price is right, customers are more likely to buy even if it is an unfamiliar way to shop.

Cable television has taught us that, at least for a major portion of the population over the age of thirty, movies in the home are more attractive than in the theater. Pay-television, either on the cable or over-the-air, has been a market winner. Of course the TV habit was formed long before television-for-pay, and neither movies nor free-TV have disappeared; people want choices.

WHY AND HOW THE INDUSTRY IS GROWING

An industry will grow if there are economic advantages to the products and services it provides and there are several such advantages for the network information services industry. Network information services use distributed data processing and this offers opportunities for sharing often complex and high-valued resources thereby saving money. Distributing computing among many processing locations and many nodes increases system reliability. Under certain conditions, distributed processing can keep communications costs down. Perhaps most important of all is that only through a distributed system can a true marketplace operate. A significant characteristic of a wired network is that the marginal cost of connecting the next subscriber on the network becomes smaller as the number of clients on the network increases. The more buyers and sellers in the marketplace the greater the value of the market to everyone. The network marketplace is, consequently, a most efficient place to do business.

The other economic advantages that drive the network information services industry follow.

The Network as a Production Facility

A large distributed network of computers, software, terminals, and databases provides an ideal production facility. New services can be manufactured by providing added value to existing services, producers can pool resources to achieve economies of scale, suppliers can specialize in processes and products in which they have a competitive edge, and there is even room for the small producer or cottage industry to custom tailor services to an individual user's specifications. This aggregation of interconnected resources provides a massive production facility which enables producers to purchase only the amount they need of each of the factors of production and to change the mix easily and quickly as production demands change. Producers may purchase information, software programs, specialized design information, consulting, and other forms of expert advice and integrate these products and services directly into their production processes. There is no need to spend valuable resources on the manufacture of intermediate products in order to produce final products. Furthermore, these products can be purchased from suppliers on the network, suppliers that increasingly will be scattered throughout the world.

Perhaps the most significant resource available to firms on the network marketplace is money in the form of information about money or what might well be called "electronic money." We shall have more to say about this new phenomenon subsequently.

The Network as a Communications Medium

The network provides access to scarce or infrequently needed resources which an individual cannot justify owning. The network provides rapid mobility of information among and between the community of individuals and institutions who make up the network. What is new and important about this medium of communication is that it provides extremely rapid transfer of symbols and images at much higher speeds than human conversation or the postal services and with greater selectivity than television and other mass media.

The Network as an Organizational System

Network information services link tasks, people, resources, and management in various configurations, as temporary or as permanent as desired. These organizations permit coordination and control of dispersed functions in ways that are most suitable to the process, people, institutions, etc. being coordinated and controlled. Multiple organization structures can coexist in the system simultaneously. For example, when one configuration of the network is used for hotel or airline reservation processing, another configuration can be used for home protection services or electronic banking.

It is worth dwelling for a moment on this role of the network. As firms seek to expand their enterprises, both horizontally and vertically across the nation and across national boundaries, they recognize the extraordinary power of computer-communications networks to extend and improve their spans of control. New technology offers them the opportunity to construct their own networks designed according to their particular operational needs. Distributed management is made more efficient via distributed networks.

Couple these features to a society of people who must work harder and longer in order to live in an inflationary economy and in which travel is becoming more expensive and burdensome, and the prospect of doing business or daily chores remotely on the network becomes quite attractive.

The elements of this industry have been around for some time but have been linked peripherally and often tenuously. There are the information or knowledge production industries, to use Machlup's phrase, whose primary products and services include research and development and information collection and creation in the fields of law, engineering, architecture, business, the humanities, and the sciences. There are the information distribution industries, the communications industries, or public information services, whose primary products and services include the delivery of education, and there are the telecommunications industries, both regulated and unregulated. Finally there is the computer industry which has provided the impetus for these industrial segments to integrate. More rapid industry integration is now occurring as merchandisers, providers of transaction-based services such as banking and brokering, recognize that their products are information products, and as government agencies producing information discover the value of the mass and specialized media for reaching out to citizens and being reached by them. The entertainment industry now recognizes that there are many audiences on the networks and that the viewer it attracts is somewhat more discriminating and will pay for individualized attention.

PREREQUISITES FOR AN INDUSTRY

First and foremost, ubiquitous, low-cost, accurate, flexible, and responsive telecommunications systems are necessary (but not sufficient) prerequisites for an expanding network services industry. The telecommunications industry is moving in this direction with the large and growing number of media channels that are becoming available. Transmission is not likely to be a barrier to the network services industry.

Second, access to these networks is vital and can be achieved via terminals. While there are several options for intelligent terminals that perform transactions on the network marketplace, their costs will depend, at least in part, on consumer demand for network services. As the demand for terminals increases and with that their production, we can expect their costs to fall.

Estimates of households with personal computers by the middle of the 1990s ranged from 30 percent to as high as 60 percent. However it is not sufficient for the home to have a personal computer no matter how intelligent. That computer must be able to communicate and even the most optimistic forecasts show that less than two personal computers in ten are likely to have communicating capability by the middle of the next decade.

The videotex terminal is another alternative, somewhat lower in cost if limited in functions. We cannot discuss the videotex terminal without taking into consideration the intelligent telephone. This instrument is very likely to be a hybrid device, somewhere between a videotex terminal and a personal computer. Its potential lies in its being a part of the telephone system family. As a member of that family, the telephone can offer access to intelligence at central offices where the high cost of high intelligence can be shared among all of the users of that office. Furthermore, a telephone can be offered to households for a monthly rental fee rather than for outright and possibly expensive purchase.

Third, there must be a demand for services. There is much we can learn about the network marketplace from early experiences in cracking the teletext/videotex markets. These experiences indicate a fringe audience for non-purposeful information seeking from the home. There is an exception to this in the relatively high percentage of newspaper readers accessing the classified ads. But this is, essentially, purposeful information seeking since more than a quarter of the searchers make purchases. As few as 3 percent of households report visits to a library in the previous year, only about 5 percent have used a dictionary during the previous month, and most people do not spend more than thirty-five seconds per day actively seeking information.

When more purposeful information seeking is examined, such as that performed on high-value, high-cost business networks, the story is quite different. These information services are designed to meet the needs of the users who have been doing their work, perhaps less productively in the past. Indeed, the relative early success of this market is evidenced by the more than fifteen million computer and computer-like terminals these information workers have at their fingertips and their willingness to access databases such as Dow Jones more than one million times a day. An important and perhaps the most significant feature of this service is that it serves the needs of the purposive information seeker by being interactive and intelligent.

A fourth prerequisite is ease of access to information and transaction services. This requires standardization of protocols and languages and the development of access procedures that are less esoteric and time consuming. A simple handshake procedure, like calling on the telephone, is required. This transparency of networks may be the most difficult prerequisite to achieve.

Finally, there must be innovative product and service design. Many "natural" information products and services exist; banking is certainly one of them as is electronic mail. Firms are recognizing that by automating the order-taking and processing procedures they can significantly reduce the number of human transactions required. Improved means for purchasing via the network will avoid the problems we encountered in buying our computer game in Chapter 8. But this takes very imaginative and clever selling; a new breed of product marketer is needed, one that can make purchasing via the network as exciting as shopping on New York's Fifth Avenue or Beverly Hills' Rodeo Drive.

Critical Issues for an Emerging Industry

The emerging network information services industry raises many critical issues that challenge a democratic society, among them, privacy, and the rights to information and to its equitable distribution.

Privacy The essence of this postindustrial period is that production and use of information is the driving force of social change rather than the production and consumption of manufactured goods. It is not surprising, therefore, that the archetypical issue of this period is the protection of individual privacy. What Cooley called a person's "right to complete immunity; to be let alone" in his *Treatise on the Law of Tort* in 1879 is hardly a new issue, but it takes on a special significance in an era of increasingly sophisticated and widespread data processing and telecommunications capabilities, more numerous and larger data banks, more government, more ingenious criminals and an ever more intricate economy. Was life any simpler 100 years ago? Consider these words of Brandeis and Warren from their 1890 *Harvard Law Review* article titled "The Right to Privacy":

> The intensity and complexity of life, attendant upon advancing civilization, have rendered necessary some retreat from the world, and man, under the refining influence of culture, has become more sensitive to publicity, so that solitude and privacy have become more essential to the individual; but modern enterprise and invention have, through invasions upon his privacy, subjected him to mental pain and distress far greater than could be inflicted by mere injury.[1]

Our concern with threats to privacy began in earnest in the 1960s with the revelations about electronic surveillance and bugging, government dossiers on dissenters, especially opponents to the Vietnam War, government proposals to establish a National Data Center linking files in some twenty federal agencies, and the first computer crimes. Much the same concern was expressed in Europe where privacy laws were quickly enacted. In the United States attempts to deal with this issue have been much slower and more fragmented. A host of official study groups explored the questions and issued a number of reports. All of them called for new legislation to remedy the defects of the major privacy-related legislation on the books, rendered ineffective or at best partially effective by the rapid emergence of the information technologies and the telecommunications-computer networks spanning the nation.

But despite pervasive public fears and growing political attention, the notion of privacy itself remains ill-defined. The Privacy Protection Study Commission did not even attempt a definition but rather spoke only of three broad "privacy concerns":[2]

• The need to protect against unwarrented intrusions via information collection practices.
• The need to ensure fairness in recordkeeping (and especially to see that inaccurate, incomplete, or obsolete information is not being used in decisions affecting individuals.
• The need to establish or preserve reasonable and enforcable expectations of confidentiality (in such relationships as those of customer-bank, client-lawyer, patient-doctor, husband-wife, etc.).

In most people's minds the concept of privacy is coupled with security and reliability. A major computer manufacturer, for example, defined "data security" as the "the protection of data from accidental or intentional disclosure to unauthorized persons and from unauthorized modifications." But much depends on where you stand. If security can be viewed as the degree to which a system or the data it processes to create information are vulnerable to penetration by unauthorized persons, whether intentionally or accidentally, then invasion of privacy is only one type of security failure, along with theft, fraud, sabotage, and pranksterism. Yet in the world of the network information services industry, systems are always interpreted broadly to include a configuration of policies, hardware, software, procedures, people, and institutional environments, all interacting toward a single objective. In this case a perfectly invulnerable system guarantees perfect privacy. It seeks to be invulnerable by embodying, among other features, safeguards to carry out a desired policy on privacy.

What is an appropriate policy?

As with all policy issues, contradictions surface to challenge our values. For example, we wish to preserve the privacy of our bank accounts, especially from casual searches by government. On the other hand, should we have reason to think that someone has been tampering with our accounts, we want "the law" to investigate immediately. We are willing to open our accounts to catch a thief.

Consider the dilemma of point-of-sale and credit transactions. Such transactions will increasingly involve instant (electronic) authorization. But this means that users can be instantly located and also can be tracked as they move around a city, country, or even the world.

Users of electronic funds transfer services often wish to receive a complete record of their financial transactions from the terminal. Indeed, one of the selling features of electronic funds transfer is access to records on demand. While this convenience allows for more efficient family accounting, it also substantially increases the ability of system operators to develop profiles on individuals.

Privacy controls over data flows between private entities is a major objective of the European countries in their desire to ensure the privacy of nationals. Many of these messages are encrypted. Will governments be forced to decode the message to determine if a citizen's privacy is being violated?

Information Equality Until the beginning of 1984, our nation prided itself on a single national telephone network that sought to provide universal service at affordable costs. In practice, if not by law, telephone service was considered a public good; those who could not afford to pay the tariffed rates paid lower rates for "lifeline" services; even families on welfare could purchase telephone services.

In the late 1970s rapidly expanding industries and businesses throughout the nation sought more diverse forms of communications, and the concept of enhanced services was developed. These enhanced services were taken out of the regulated marketplace. Increasingly our nation is being served by a diversified, fragmented, multinetwork communications system in which rates are set by the marketplace rather than by federal or state tariff bodies. The number of special purpose networks providing digital, voice, image, and video services is increasing almost geometrically every month. Indeed, the demands for private lines have overwhelmed and overburdened the carriers.

These networks are often service- and industry-specific in order to achieve the coordination and management efficiencies and the resulting economies noted above. Many Fortune 500 companies and about a third of the medium-sized industries in the nation already account for almost 80 percent of the traffic on these networks.

Consumers and smaller businesses find the network marketplace too expensive. While savings may be available once you are on a network, the high cost of entry may be beyond the reach of many pocketbooks. Furthermore, the networks are not easy to operate; they are not transparent and trained operators on special terminals may be required to accessed information. Some observers see a situation emerging reminiscent of the early days of the telephone with many incompatible networks designed to deliver services to those who can afford the cost of entry and operation.

That there is a public good aspect to the telephone is undeniable; however, it is difficult to prove and sell to marketplace afficionados. There is considerable evidence that the cost of a telephone call does not at all reflect the value received from that call. Consider those telephone uses that are performed on behalf of community organizations, neighborhood groups, and political parties. Telephone use is a differentiable product whose prices should more fairly reflect the value of the telephone call to the user. And, in some cases, the value may accrue not to the individual caller but to the community at large.

In the early days of the telephone, the significant issue was equality of access to communications; today, however, the issue has become equality of access to information. Some have argued that there is the danger of an information elite emerging and that there will be two classes of people and businesses, those that can afford to access information—the information rich—and those that cannot afford to access information—the information poor.

Should the network information services industry evolve as other industries have, there is likely to be a degree of market segmentation. This will result in pricing strategies designed to meet several levels of market demand and abilities to pay. We must, however, be especially vigilant to see that an inequitable pricing structure does not emerge. We are dealing in unusual products—information and knowledge. The network information services industry is essentially an *information* industry and information has important social as well as economic values.

THE ECONOMICS OF THE NETWORK MARKETPLACE

In Chapter 7 on intelligent telephones we found that it is possible to communicate with people across time and space. The intelligent telephone is attractive because it can make life more convenient and business more effective by offering alternative behavior options; we can choose to travel or choose to "telecommunicate."

Network information services offer the additional prospect of significant economic benefits; they are a means of reducing transaction costs by the substitution of intelligent technology for intelligent people. For example, the sixfold reduction in the cost of a bank transaction is achieved by eliminating the check-processing procedure and substituting an intelligent machine on an intelligent network—the automatic teller machine—and calling on the customer to perform as instructed by the machine. Indeed, several major banks have attempted to "encourage" customers whose account balances are below a particular amount to use automated tellers, reserving human tellers for the bank's more prosperous customers. So far these efforts have not been too successful; competitors have reacted by promoting the opportunities their customers have to talk to "real live people."

The network marketplace is nonpolluting and energy efficient. The network can create a climate in which diversity and pluralism can flourish; network businesses are less capital intensive since they do not require physical facilities. This is especially important in an economy which is likely to see an inflation rate of between 7 and 10 percent in the foreseeable future. With a tightening of the money supply that is likely to follow, network businesses that can be launched without heavy investments—businesses that require access to a network, software and databases, and relatively inexpensive terminals rather than expensive transportation systems and real estate—can be very attractive. Network businesses are less labor intensive, but call for new skills that require continuing education, job upgrading, and job retraining.

The information era is characterized by the emergence of information industries and businesses, activities that lend themselves to the network marketplace. As we saw in the previous chapter there are increasing demands for information as bureaucracies, both in private enterprise and government, shore up their economic positions. Increasing labor costs stimulate the search for greater productivity and computers communicating with computers on networks appear to satisfy this demand. The worldwide production and distribution of information is creating increasing foreign competition. As networks proliferate over satellites and cables, there are strong incentives to take advantage of the international distribution of labor and capital. At the same time, offshore manufacturing becomes more economically viable because it is easy to control and manage these facilities remotely via the network.

The network marketplace is the information economy with all of the challenges and promises to which a new era exposes us. We know that new markets must be found, that information markets are likely to be highly fractionalized and specialized. The capture of these markets requires that we obtain information about them. The network marketplace with its multiplicity of information sources is one way we can do so.

THE HUMAN SIDE OF THE NETWORK MARKETPLACE

How often have we heard and said that computers, terminals, recorded voices, and computer-simulated speech are depersonalizing? Then along comes the network marketplace with its promises to reduce the human content in our transactions. Resentment and fear are already prevalent among many who have argued with a computer over an incorrect bill, and who have had human but very impersonal clerks report that "the computer made an error" while they wait for that error to be corrected. Acceptance of many transaction services such as remote purchasing and banking has always fallen behind expectations. A major reason is that so many of these systems are complex and error prone; once burned, a client is likely to be cautious about trying again. Any significant breakdown in privacy safeguards results in an increased fear of the technology and raises the specter of the individual's inability to control his or her environment.

Then too, there is the desire for choice. It matters little to many consumers that, in an effort to satisfy all customers, the bank and the merchandiser must provide dual transaction channels—via the network and over-the-counter—and if one provider's costs rise, there is always another provider whose costs may not have risen or who is prepared to offer the same services at lower costs.

There are, however, flaws in this procompetitive argument. For one, networks favor large service providers who are increasingly absorbing the smaller providers, thereby eliminating options. There are also the growing numbers of elderly who are likely to find adapting to a machine most difficult. And what about the information poor without access to terminals and networks or to the information necessary to deal with the marketplace? Are they to be left out of the information economy?

Despite these concerns, the nation's demographics seem to be moving in a direction that favors the growth of network information services. Single-person households, smaller families with two working parents, and the growing independence of women create a need for more efficient use of time. The network marketplace offers an option that can reduce the time-consuming activities of banking, some shopping, and general information seeking. With the need to share household duties, a network marketplace that is accessible at odd hours can be very attractive to young families that have the resources to purchase the terminals and services necessary to make use of the network.

Finally, the network can expand group interaction and create social networks not bound by geography, but by areas of interest. An accessible network marketplace could create a climate in which diversity and pluralism can flourish.

REFERENCES

1 Brandeis, Louis D., and Samuel Warren, "The Right to Privacy,"*Harvard Law Review*, Vol 4, 1890, p.196.
2 U.S. Privacy Protection Study Commission,*Personal Privacy in an Information Society*, Final Report (Washington DC.: U.S. Government Printing Office, July, 1977), p.14.

ADDITIONAL READINGS

Because the notion of network information services and the network marketplace is a relatively recent one, the literature is, as you might guess, quite scattered across many fields. Here is a selection that tries to cut across these fields to provide a broad picture of that notion.

Dordick, Herbert S., Helen G. Bradley, and Burt Nanus, *The Emerging Network Marketplace* 2nd ed. (Norwood, NJ: Ablex Publishing Corporation, 1985). This is the first book that expounded the notion of NIS and network marketplace. In particular, Part V on key issues discusses many of the ideas found in this chapter.

Moss, Mitchell L., ed., *Telecommunications and Productivity* (Reading, MA: Addison Wesley Publishers, 1981). A collection of articles that tackles the difficult issue of productivity in the information era.

Schiller, Herbert I., *Who Knows? Information in the Age of the Fortune 500* (Norwood, NJ: Ablex Publishing Corporation, 1981). An excellent critical view of the information society.

Toffler, Alvin, *The Third Wave* (New York: William Morrow and Company, 1980).

Wicklein, John, *Electronic Nightmare* (Boston: Beacon Press, 1979). A rather frightening view of the network marketplace; is it realistic?

PART **SIX**

TRACKING THE FUTURE

Up until now, we have resisted a strong temptation to speculate about the future. However, that seems de rigueur for texts on telecommunications technology. We continue to resist this temptation, at least in part. We shall not speculate about the future of technology; rather we shall inquire into how an information society is likely to emerge. We turn now to an examination of how society will adapt to the technologies we have been discussing or, preferably, how society may alter these technologies to meet human needs and behavior.

CHAPTER **14**

SPECULATIONS ABOUT AN INFORMATION SOCIETY

We do not attempt to offer a prediction of the future, for that requires the gift of prophecy, with which few of us are blessed. Prophets are doomed whatever their prophecy; if they are wrong they are forever discredited as unreliable and not worth attending to. To paraphrase Gilbert Chesterton,

> The prophets took something or other that was certainly going on in their time and then said it would go on more and more until something extraordinary happened. And very often they added that in some odd place that extraordinary thing had happened and that it shows the signs of the times.... The players (in the Cheat the Prophet Game) listen very carefully and respectfully to all that the clever men have to say about what is to happen in the next generation. The players then wait until all the clever men are dead, and bury them nicely. They then go and do something else.

If the prophets are right they are likely to be stoned to death should their prophecies prove unpopular and we cannot do anything about them. Remember John O'Hara's *Appointment in Samarra* in which Death tells the merchant whose servant, having seen Death in the morning, had borrowed the merchant's horse to escape to Samarra, "I was astonished to see him in Baghdad for I had an appointment with him tonight in Samarra."

Less than twenty years ago it was risky if not foolish to forecast the network marketplace. Today that marketplace is a reality, an emerging industry with its own financial, organizational, production, marketing, and regulatory practices and regimes. While there are barriers yet to overcome, there is optimism that the market will accept the network services industry with sufficiently open arms and pocketbooks to make it a worthwhile and long-lived endeavor. There is little question that the forces directing the technological futures of the network marketplace are in place but we might well question the optimism of the marketers whose forecasts depend on the shifting attitudes and often uncertain behavior of people. Man is, after all, not as malleable as a semiconductor or an optical fiber and makes choices that are not always rational, despite the fond wishes of economists. Nevertheless, Herbert Simon is probably correct when he argues that "In recorded history there have perhaps been three pulses of change powerful enough to alter Man in basic ways. The introduction of agriculture... The Industrial Revolution... (and) the revolution in information processing technology of the computer...."[1]

We shall explore several societal futures that some fearless forecasters have put forth and inquire into their reasonableness in the light of human behavior and history. History is, after all, our best forecaster of the future, even if it is often overlooked by futurists.

YONEJI MASUDA'S COMPUTOPIA

Computopia, the computer utopia of the Japanese plan for an information society, is a society that brings about a general flourishing state of human intellectual creativity instead of affluent material consumption. It is a society where people may "draw future designs on an invisible canvas and pursue and realize individual lives worth living." It is Butler's Erehwon made possible by the remarkable powers of the computer and the availability of the information and communications technologies.

Masuda's information society is a new type of human society, vastly different from our industrial society. Information machines and telecommunications—intellectual power rather than steam engines and mechanical power—are the core technologies to enhance the human being's intellectual rather than physical abilities. Information utilities will replace machinery-based factories and there will be a world market for information. Knowledge frontiers, not land frontiers, will be the targets for expansion. Private enterprises will be replaced by enlightened public enterprises in which volunteerism and volunteer communities will assume leadership. Parliamentary democracies will give way to participatory democracies driven by citizen movements that are devoid of narrow special interests.

The core of this vision lies in the replacement of humanity's need for material goods by a new orientation toward the "pursuit of a self-determined goal." The productive power of societies will be directed toward the development of information and human needs will be satisfied through that development.

How is this brave new world to be achieved? The plan of the Japanese Computer Usage Development Institute envisions a series of telecommunications and information projects that span a wide range of social and public activities, leaving to industry and the international trade marketplace the task of creating the tools required for this societal transformation. These include an administrative data bank which provides the raw material from which value-added information products can be developed, a plan for a computopolis, a "computerized city" with a broadband telecommunications infrastructure of optical fibers and coaxial cable sufficient to provide interactive data and visual communications, and several regional health and education facilities which will experiment with, demonstrate, and provide telemedical and teleeducational services.

Education and training of information workers will require a substantial public investment to which business and industry will contribute, for their productivity depends on the availability of workers trained for the new information jobs this information society will require.

Computopia will, of course, have its problems; privacy and social control through information raise ugly specters and the loss of traditional jobs will cause shocks that may endanger social stability. Service jobs will replace manufacturing jobs but for Masuda this appears not to create any special terrors. In many Eastern societies and especially in Japan the individual is not degraded by performing a service and producing a "good" that is ephemeral. Japanese executives and professionals find personal satisfaction and gather public appreciation by regularly performing religious and family rituals that in the West are relegated to holidays, rare evenings at home, and short weekends.

This vision is reminiscent of the utopias that have enriched the literature of many nations and whose promises brought so many immigrants to the New World. While the origins of these utopian movements differed—some sought to escape from religious persecution, others from political oppression—they all seemed to establish a new set of rules and procedures that challenged human behavior. The dreamers and founders never understood why others could not see the reasonableness of their utopian proposals and adapt as well as they did.

Technological utopian visions are, of course, technologically deterministic; they assume that human behavior is easily changed by the rational actor recognizing the extraordinary value and benefits of the technology. They assume a one-way path from technology adoption to behavior modification. As we have so frequently noted in the previous chapters, the telecommunications technologies are unique in that they offer users so much choice with the ultimate choice being to use or not to use them. While the automobile replaced public transportation in many cities, a new communications medium did not replace existing media, thus resulting in the wonderful consequence of allowing us so many media choices. How we make these choices is defined by our communications behavior, and our behavior is shaped by our attitudes toward the media and technology.

An analysis of attitudes toward the media reveals many layers of interrelated and increasing complexities. Despite the many years of research into communications behavior, we still find many unknowns, perhaps because as new media are available we have more from which to choose, thereby creating additional measures with which we seek to make choices. Williams, Rice, and Dordick have noted that "choice of medium and style are complex additudinal considerations and although this idea has been with us for nearly all of the history of commentaries upon communication, it is still being overlooked in many contemporary studies of the new technologies."[2]

Is there a form of "media stereotyping" which creates attitudes toward media? We know that attitudes toward different modes of communication—newspapers, broadcast and cable television, videocassette machines, telephones, radio, and computers—are multidimensional and complex depending on familiarity, personalness, importance, ease of use, and quality of use, among other factors. We also know that place has an important impact on the individual's use of a particular medium; how and for what purpose he or she uses the telephone is determined by whether that phone is in the home, in a public telephone booth, or at work. With the merging of home and office as the technology of communications mobility seems to foretell and with the erasing of images of time and space that we can expect from computer communications and modern telephone services, more uncertainty about human communications behavior and about how we will use these wonderfully flexible media mixes is likely to surface.

Communications technology does not drive communications behavior, rather there is a symbiotic relationship between the two. This relationship will determine if the future is Masuda's Computopia and not simply the computer and telecommunications technologies.

MACHLUP, BELL, PORAT, AND THE NETWORK MARKETPLACE

Fritz Machlup envisioned an economically driven information society, one that would transform our way of life through jobs, industries, and trade. The production and distribution of knowledge would create information work requiring new kinds of activities in new kinds of industries. Knowledge would thus become a form of technology itself—an "intellectual technology" to quote Daniel Bell—and information the basic resource in this postindustrial society. The growth of knowledge work and knowledge workers is one of the major shifts in contemporary society and as such is a major force for social change.

With information as the central capital—indeed, the cost center and crucial resource in our economy—a major restructuring of our economic base will take place. Porat confirmed that this was, indeed, occurring in his 1977 report indicating that by 1967 about one-fourth of the U.S. Gross National Product reflected the production, processing, or distribution of information goods and services. With an additional 21 percent of the GNP devoted to satisfying information demands and to providing the coordination a complex society with its private and public bureaucracies requires, Porat suggests that more than 46 percent of the workforce earning more than 53 percent of the income falls into the category of information workers.

The economist's vision of an information society is one in which efficiency rules and, hence, human behavior adjusts. Information work is intellectual work and with computers and telecommunications as the tools of that work, we will certainly become ever more efficient. If less work is required to produce today's products and services, we shall invent new products and services for worldwide markets, for the distribution of information products and services will take place on telecommunications networks. The efficient worldwide distribution of labor will ensure the proper allocation of competitive advantages; in the long run, every nation and every person will find their appropriate place in the scheme of things.

This is the vision of the network marketplace, and it is far from problem-free. While economists love to explain away today's problems by looking "to the long run," this shift of our nation's production base is bound to result in the dislocations and value clashes we discussed in the previous chapter.

To some, the network marketplace vision of the information society represents a gloomy future where even ideas in the form of information have been priced, and perhaps priced out of the reach of many in society. Jacques Ellul makes this point when he comments that we have "technologized" ourselves at the expense of traditional human values. To which Bell argues that we have the opportunity to reinforce these values since we shall no longer emphasize getting more with less or sheer productivity and we shall be more concerned with social consequences and objectives. Bell defines an "economizing mode" as one oriented toward functional efficiency and the management of things (and people are treated as things). He defines a "socializing mode" as one that establishes broader social criteria, but which, necessarily, involves the loss of efficiency, the reduction of production, and other costs that follow the introduction of non-economic values.

It is not at all clear that social equity must necessarily lead to economic inefficiency. Examples abound where social equality, reaching for the "socialogizing" goals, and efficiency can coexist. Economists measure efficiency, in part, by the amount of "capital formation" or savings a society generates. Consider, then, that the nations with the highest postwar savings rate include Austria, West Germany, the Netherlands, France, and Italy, nations that have generous social welfare programs. Another argument is that a progressive tax system is bad for economic growth. Yet the nation with the stiffest rates of effective corporate tax, Japan, has had enviable records of productivity and economic growth for many years running.

If there is to be a postindustrial information society that avoids the failures of the industrial society we shall have to carefully balance values in praise of efficiency with those that seek to preserve equity. We perceive a convergence of Masuda's future with that of the postindustrial network marketplace.

GERSHUNY'S SELF-SERVICE ECONOMY

The need to create futures derives from observations that there are wrongs in our society and that technology offers some means to redress these wrongs. What are the problems we believe the technologies we have been studying can make right?

1 Our society is subject to scarcity; all of society's members desire the availability of more goods to meet their basic needs for food, shelter, and clothing.

2 Consequently, there will be a tendency to exploit economies of scale to the fullest possible extent. Labor will be distributed to produce more goods to meet basic demands.

3 New technologies are scarce, especially those that can be exploited to obtain the benefits of economies of scale.

4 In particular, communications are sparsely provided. This poses problems for the administration and control of enterprises. This shortage also limits the forms which organizations can take.

Increasing wealth means less unmet need for goods to satisfy basic needs. And in any case, the visions of the information society we have been discussing argue that material needs will be replaced by information needs leading to the satisfaction of intellectual needs.

In our discussion of the marriage of telecommunications and computing, we noted that their very nature creates economies of scale that have not yet been exhausted. Indeed, exhausting these economies is unlikely in the foreseeable future. We are developing a technological base that is adaptable to small and large productive units and to small and large needs, that is both small scale and large scale, and that presents the human face Schumacher dreamed of for his small and beautiful technologies.

Technologies may no longer be scarce. Developments in modern telecommunications are wiping out shortages of communications, unless, of course, a shortage is created through inequitable pricing mechanisms. Instead of social shortcomings that demand technical solutions, we now often find multiple technologies chasing scarce applications. Teletext and videotex have been called technologies in search of markets. This quality is true not only of the information technologies; Gershuny identifies 178 distinct unconventional passenger transport systems, all in some stage of research or development, only a tiny fraction of which will ever be evaluated at full scale, and none of which might ever come into general use.

Gershuny argues that we are not entering a new era but rather continuing along the paths of technological innovation set for us by the industrial revolution. Consumption patterns are not radically changing; indeed, the demand for goods is increasing more rapidly than the demand for services and information. If we were to review the statistics compiled by Porat and rather than distinguish between primary and secondary information workers and output assign the latter to the manufacturing sectors, the output of the manufacturing sector would grow more rapidly and to ever higher levels than the output of the information workers and their industries.

Information work is often personal work. Information activities are often self-service activities; we sit at our own personal computers, personal word processors, and other information terminals. Information work is not production line work, it is intellectual piecework, and as such, it is suitable for the cottage as Toffler points out.

However, increased capital investments must be made in order to perform information work. It is claimed that the capital investments now being made for office personnel rival those that have been made in the past in the factories, a sharp rise from the average $3000 to $25,000 for a word processor or other sophisticated information terminal.

Increased expenditures for capital goods are made by households when it becomes more economic to do for yourself than to have others do for you. Instead of hiring higher-priced domestic help for cooking and washing in the home, we purchase mechanized and computerized small appliances, dishwashers, and ovens. In most industrial nations transport services are being replaced by transport goods, private automobiles replace buses and trains because we place a higher value on traveling by ourselves and controlling our mobility than on using public transport.

This do-it-yourself trend also is moving us toward entertainment at home rather than in cinemas or theaters. But to enjoy this entertainment at home, we must make capital investments in cable television and videocassette recorders. To capture the benefits of the telecommunications-transportation tradeoff we purchase personal computers and add modems which enable us to bank and shop from home.

Gershuny's vision is that of a self-service society which is merely an extension of the industrial revolution with a greater mix of work options. Instead of working in factories on long production lines, we work alone on our individual intellectual piecework which is eventually assembled on the information network marketplace and sold in varieties of job lots! It is a vision of a highly isolated way of life, where private interests are valued more highly than public action. Is this a historically valid form of behavior? Let's see.

Private versus Public Actions in the Information Society

In his monograph entitled *Shifting Involvements,* Albert Hirschman examines society's swings between public actions and private interests, swings which he attributes to a dialectic of frustration. History is our best guide to the future and Hirschman's short trip into the history of societal frustration may provide a guide to the future of the information society.

Hirschman argues that disappointments in private actions can lead to frustration and boredom, and that the lack of private actions can lead to the desire to experience the pleasures of public actions. Conversely, should public actions lead to disappointments, frustration, and boredom, there may be a return to purely private goals and actions.

Pendular movements in society are common, Hirschman argues; it is possible to discern a cycle between conservatism and liberalism, a cycle of perhaps every fifteen to twenty years. There seem to be similar cycles between a society's desire to be publicly active and privately contemplative.

We seem now to be in a contemplative mode, perhaps because we were disappointed with the activism of the '60s and '70s. This trend toward privateness is being reinforced, catered to, and encouraged by the network marketplace. If this is a cyclical movement, how long will it last? Society more often appears to be in a contemplative mode and bursts forth periodically into a public mode for only short periods of time. This may be good news for the proponents of the information society, a society that supports the vita contempletiva. They would argue that while we may see a brief surge of desire once more to attend theaters, do our banking in banks with "real" people, shop in supermarkets and supermalls, talk to people rather than leave messages, work in large offices, even side-by-side as on an information production line, we shall return for another long period to our contemplative mode.

TELECOMMUNICATIONS AND THE TRANSFORMATION OF THE WORLD'S BUSINESS

Over the next two decades we can expect to see the rapid diffusion of the telecommunications technologies we have been studying. Optical fibers will span the globe competing with satellites which will have the capacity to serve earth locations only miles apart; increasingly intelligent networks and terminals will operate over essentially cost-free wired and broadcast transmission systems, made so by even greater utilization of computer technologies in telecommunications; and worldwide instantaneous electronic delivery of information will take place in multiple modes: voice, images, data, and video. A new form of global infrastructure is emerging, one that provides both communications and computing and is intelligent.

In terms of output and employment, the telecommunications industry is relatively small when compared to other industries. Nevertheless, it greatly influences the management styles and business growth opportunities of other industries. Telecommunications can greatly expand the control and coordinating capabilities of management and, thereby, play an important role in shaping industrial structures. It is an industry poised on the brink of enormous technological advances, capable of greatly increasing the productivity of other industries as well as its own. There appears to be no limit in the foreseeable future to the flexibility and diversity of telecommunications services, nor to the reduction in transmission costs that will continue to make communicating an attractive and important management activity. The clear implication is that a shift in office/information tasks from labor intensity to capital intensity will increase efficiency as well as reduce the number of employees required.

The improved quality and reduced cost of telecommunications services—intelligent telecommunication services resulting from the diffusion of new generations of equipment and systems—will continue to have an impact on high technology manufacturing industries and on the service industries, both of which are dependent on the transfer of information. Finance, insurance, brokering, information processing, and the news media among others are rapidly becoming computerized because of their need to transfer large volumes of information across nations and continents at costs that are far below the value of the information being transferred.

Since the 1970s several important trends have become apparent in the telecommunications industry and we can expect them to continue, indeed, to accelerate in impact:

1 The influence of service providers over the telecommunications equipment market will continue to decline. This is largely because new products and technologies are leading to rapid growth in private equipment markets and new export markets are emerging. As telecommunications equipment becomes more computerized, computer and computer terminal firms are entering the telecommunications equipment market.

2 New technologies have encouraged firms not in the telecommunications market but dependent on the transfer of information between their branches both at home and abroad (financial service firms, in particular) to enter the telecommunications equipment and services markets.

3 As firms seek to expand their enterprises, both horizontally and vertically across national boundaries, they recognize the extraordinary power of computer-communications networks to extend and improve their spans of control. New technology offers them the opportunity to construct their own networks designed according to their particular needs. This is the trend away from nationalization of telecommunications infrastructures and toward the conversion of national PT&Ts to private companies and the introduction of competition where regulated monopoly previously was the rule. It has already occurred in the United States; the divestiture of AT&T can be seen as a form of greater private sector involvement.

This pattern of events is being repeated throughout the developed and newly industrializing worlds. In those nations where planning for transmission infrastructures is only now under way—Indonesia and Thailand for example—the pressures from user groups, more often than not multinational users, will be responsible for forcing a shift to privately owned telephone systems. Where infrastructures already exist, as in Malaysia and the Philippines, the choice of future technology—ISDN monopoly versus a confederation of user-operated telecommunication systems that can interconnect with a public system to provide subscriber services—will be debated. Japan is already well along the path toward the creation of multiple networks serving multiple business and consumer users.

The need for improved transmission services capable of meeting the telecommunications and data communications needs of the emerging high technology industries and businesses is certainly a major force for change. But another which may prove to be a far greater force than simply the convergence of the computer and communications technologies is the desire of host country and multinational firms to enter into new forms of businesses made possible by the flow of "electronic money" and to expand these enterprises beyond national borders making use of the improved coordination and management capabilities their computer-communications networks offer them.

In the forces that are creating the information society we often overlook information itself, instead concentrating on the technologies that process and transport this information. As we noted in the previous chapter, it is wise to remember that money is information on today's international telecommunications networks, and this "electronic money" will be a major force in creating new enterprise structures throughout the world. Firms can utilize financial resources in ways heretofore impossible or difficult. This ability for intra- and inter-firm transactions opens the door for configurations of new business activities which will effectively break down the traditional barriers of product and process knowledge and experience and skills normally seen as restricting business expansion. The availability of a coherent worldwide system for handling information may explain the revolutionary dimensions emerging in world business and may well result in the information society.

REFERENCES

1 Simon, Herbert A., "The Consequences of Computers for Centralization and Decentralization" in Michael L. Dertouzos and Joel Moses, eds., *The Computer Age: A Twenty-Year View* (Cambridge, MA: MIT Press, 1980) pp. 212-228.
2 Williams, Frederick, Ronald Rice, and Herbert S. Dordick, "Behavioral Impacts in the Information Age" in Brent Rubin, ed., *Information and Behavior, Vol. 1* (New Brunswick, NJ: Transaction Books & Periodicals, forthcoming).

ADDITIONAL READINGS

As you might expect the literature on the multiple visions of the information society is large and growing ever larger. The curious and intelligent observer of the scene must, however, read beyond the information society. Indeed, to understand today's discussions, it is necessary to touch base with the industrial revolution, to see what historians believe went on before and immediately after. One ought also to consider writings on the philosophy of technology. Here is a selection of readings worth exploring.

Bell, Daniel, *The Coming of Postindustrial Society: A Venture in Social Forecasting* (New York: Basic Books, Inc., 1976). A tour-de-force in speculation that should be tackled.
Braudel, Fernand, *Civilization and Capitalism, 15th–18th Century,* 3 volumes (New York: Harper and Row, 1981–1984). This remarkable work is now available in paperback and should be in your library.
Ellul, Jacques, *The Technological Society* (New York: Vintage Books, 1964). A critical French observer views technology and society.
Gershuny, Jonathan, *After Industrial Society: The Emerging Self-Service Economy* (Atlantic Highlands, NJ: Humanities Press, 1978). An unorthodox view of the information society from Great Britain.

Hirschman, Albert O., *Shifting Involvements: Private Interest and Public Action* (Princeton, NJ: Princeton University Press, 1982).

Masuda, Yoneji, *The Information Society As Postindustrial Society* (Bethesda, MD: World Future Society, 1981). The Japanese view of an information society.

Mumford, Lewis, *The Myth of the Machine* (Boston: Houghton Mifflin, 1967). An early and important book on technology which sheds surprising insights into what is going on today.

Schiller, Herbert I., *Who Knows? Information in the Age of the Fortune 500* (Norwood, NJ: Ablex Publishing Corporation, 1981).

Toffler, Alvin, *The Third Wave* (New York: William Morrow & Company, 1980). A journalist's view of an information society.

GLOSSARY OF TERMS

AM (amplitude modulation) one of several ways of modifying a radio or high frequency wave to carry information; the AM radio band is that portion of the radio spectrum allocated to AM radio services. (See FM, modulation.)
analog the representation of information by continuous wave forms that vary as the source varies; a method of transmission by this method. Representations which bear some physical relationship to the original quantity, usually a continuous representation as compared to digital representation. (See digital.)
antenna(e) a device used to collect and/or radiate radio energy.
artificial intelligence computer programs that perform functions, often by imitation, usually associated with human reasoning and learning.
ASCII (pronounced "ask-ee") American Standard Code for Information Interchange. The binary transmission code used by most teletypewriters and display terminals.
band a range of radio frequencies within prescribed limits of the radio frequency spectrum.
bandwidth the difference in hertz (cycles per second) between the highest and lowest frequencies in a signal or needed for transmission of a given signal. The width of an electronic transmission path or circuit in terms of the range of frequencies it can pass. A voice channel typically has a bandwidth of 4000 cycles per second (or hertz); a TV channel requires about 6 million cycles per second (megahertz).
baud bits per second (bps) in binary (two-state) telecommunications transmission. After Emile Baudot, the inventor of the asynchronous telegraph printer.
binary a numbering system having only two numbers, typically 0 and 1.
bit (binary digit) the smallest part of information with equally likely values or states, "0" or "1," or "yes" or "no." In an electrical communication system a bit is typically represented by the presence or absence of a pulse.
bps bits per second.

broadband a term often used to describe a range of frequencies wider than that required for just voice communications. Also a term used to describe systems and equipment with wide bandwidth that can carry these ranges of frequency. (See narrowband.)

broadband carriers the term used to describe high capacity transmission systems used to carry large blocks of telephone channels or one or more video channels. Such broadband systems may be provided by coaxial cable or microwave radio systems.

broadband channels transmission channels used primarily for the delivery of video or for high-speed data transmission.

broadband communication a communications system with a bandwidth greater than voice bandwidth. Cable television is a broadband communication system with a bandwidth usually from 5MHz to 450MHz.

broadcasting transmission intended for reception by the general public. (See point-to-point transmission.) A radio communications service for the general public; includes sound, television, data, and graphic/text transmission.

byte a group of bits operating together. Bytes are often grouped into 8-bit groups but 7-, 9-, and 16-bit bytes are not uncommon. Today's fast computers often use 32-bit bytes.

cable television (CATV) a transmission system that distributes broadcast television signals, satellite signals, original programming, and other signals by means of coaxial cable. Cable television has also been called community antenna television, cable communications, and broadband communications.

carrier a signal with known characteristics—frequency, amplitude, and phase—that is altered or modulated in order to carry information. Changes in the carrier are interpreted as information.

CCITT Consultative Committee for International Telephone and Telegraph, an arm of the International Telecommunications Union (ITU), which establishes voluntary standards for telephone and telegraph interconnection.

central office the local switch for a telephone system sometimes referred to as the wire center or class 5 office.

channel a segment of bandwidth which is used to establish a communications link. A channel is defined by its bandwidth; a television channel, for example, has a bandwidth of 6MHz, and a voice channel about 4000Hz.

channel capacity the maximum number of channels that a cable system or some other delivery system can carry simultaneously.

chip the colloquial name for the semiconductor element that has become the heart of modern computing and increasingly of modern telecommunications. The silicon base on which the electronic components are implanted is smaller than a fingernail and looks like a chip. Increasingly the chip refers to a single device made up of transistors, diodes, and other components interconnected by chemical processes and forming the basic element of microprocessors.

circuit switching the process by which a physical interconnection is made between two circuits or channels. (See packet switching and message switching.)

coaxial cable a metal cable consisting of a conductor surrounded by another conductor in the form of a tube which can carry broadband signals by guiding high-frequency electromagnetic radiation.

common carrier an entity that provides transmission services to the public at nondiscriminatory rates and which exercises no control of the message content. An organization licensed by the Federal Communications Commission (FCC) and/or by various state public utility commissions to provide communication services to all users at established and stated prices.

Comsat Communications Satellite Corporation. A private corporation authorized by the Communications Satellite Act of 1962 to represent the United States in international satellite communications and to operate domestic and international satellites.

CPU central processing unit; the component in a stored-program digital computer which performs arithmetic, logic, and control functions.

CRT cathode ray tube, a video display vacuum tube used in television sets and display computer terminals. Today, a CRT is the display terminal for video games, word processors, personal computers, and a host of other electronic information products.

dedicated lines telephone lines leased for a specific term between specific points on a network usually to provide certain special services not otherwise available on the regular or public-switched network.

demodulate a process by which information is recovered from a carrier. (See modulate.)

digital a function which operates in discrete steps as contrasted to a continuous or analog function. Digital computers manipulate numbers encoded into binary (on/off) forms, while analog computers sum continuously varying forms. Digital communications is the transmission of information using discontinuous, discrete electrical or electromagnetic signals which change in frequency, polarity, or amplitude. Analog intelligence may be encoded for transmission on digital communications systems. (See pulse code modulation.)

direct broadcast satellite (DBS) a system in which signals are transmitted directly to a receiving antenna on a subscriber's home or office via satellite rather than redistributed via cable or an intermediary terrestrial broadcast station.

downlink an antenna designed to receive signals from a communications satellite. (See uplink.)

downstream the direction of transmission of the cable signal from the headend to the subscribers. (See upstream.)

earth station a communications station on the surface of the earth used to communicate with a satellite.

electronic mail the delivery of correspondence including graphics by electronic means usually involving the interconnection of computers, word processors, or facsimile equipment.

FCC (Federal Communications Commission) the federal agency established in 1934 with broad regulatory power over electronic media including radio, television, and cable television.

facsimile the transmission of images of original documents including print and graphics by electronic means, usually via the telephone network but also by radio.

final mile the communications system required to get from the earth station to where the information or program is to be received and used. Terrestrial broadcasting from local stations, the public-switched telephone network, and/or cable television systems provide the final mile for today's satellite networks.

firmware instructions for operation of computers or communications equipment embedded in the electrical design of a microchip.

FM (frequency modulation) one of several ways of modifying a radio or high frequency wave to carry information. The FM radio band is that portion of the radio spectrum allocated to FM radio services.

frequency   the number of recurrences of a phenomenon during a specified period of time. Electrical frequency is expressed in hertz, equivalent to cycles per second.

frequency spectrum   a term describing a range of frequencies of electromagnetic waves in radio terms; the range of frequencies useful for radio communications, from about 10 kilohertz to 3000 gigahertz.

geostationary satellite   a satellite with a circular orbit 22,300 miles in space which lies in the plane of the earth's equator and which has the same rotation period as that of the earth. Thus the satellite appears to be stationary when viewed from the earth.

gigahertz (GHz)   billion cycles per second.

hardware   the electrical and mechanical equipment used in telecommunications and computer systems.

headend   the electronic control center of the cable television system where incoming signals are amplified, filtered, or converted as necessary. The headend is usually located at or near the antenna site.

hertz (Hz)   the frequency of an electric or electromagnetic wave in cycles per second; named after Heinrich Hertz who detected such waves in 1883.

institutional loop   a separate cable in a CATV system designed to serve public institutions and/or businesses usually with two-way video data and voice services.

ITFS (instructional television fixed service)   a band of microwave video channels reserved for educational and noncommercial uses.

kilohertz (KHz)   thousand cycles per second.

local loop   the wire pair which extends from a telephone central office to a telephone instrument. The coaxial cable in a broadband or CATV system which passes by each building or residence on a street and connects with the trunk cable at a neighborhood node is often called the subscriber loop.

low power television stations (LPTV)   a television broadcast station which is licensed to operate with limited radiated power and serve an area considerably smaller than that served by the full-power television station.

LSI (large scale integration)   single integrated circuits which contain more than 100 logic circuits.

megahertz (MHz)   million cycles per second.

memory   one of the three basic components of the central processing unit (CPU) of a computer. It stores information for future use.

message switching   a computer-based switching technique that transfers messages between points not directly connected. The system receives messages, stores them in queues or waiting lines for each destination point, and retransmits them when a receiving facility becomes available.

microchip   electronic circuit with multiple solid-state devices engraved through photolithographic or microbeam processes on one substrate. (See microcomputer, microprocessor.)

microcomputer   a set of microchips which can perform all of the functions of a digital stored-program computer.

microprocessor   a microchip which performs the logic functions of a digital computer.

microwave   the short wavelengths from 1GHz to 30GHz used for radio, television, and satellite systems.

minicomputer   in general, a stationary computer that has more computer power than a microcomputer but less than a large "mainframe" computer.

modem (modulator-demodulator)   a device that converts digital pulses to analog tones and vice versa for transmission of data over telephone circuits.

modulation a process of modifying the characteristics of a propagating signal such as a carrier so that it represents the instantaneous changes of another signal. The carrier wave can change its amplitude (AM), its frequency (FM), its phase, or its duration (pulse code modulation), or combinations of these.

multiplexing a process of combining two or more signals from separate sources into a single signal for sending on a transmission system from which the original signals may be recovered.

multipoint distribution service (MDS) stations intended to provide one-way microwave radio transmission in an omnidirectional pattern from a stationary transmitter to multiple fixed-point location receivers. MDS (single channel) and MMDS (multi-channel) are the means for providing pay-television programming to households.

narrowband communications a communication system capable of carrying only voice or relatively slow-speed computer signals.

optical fiber a thin, flexible glass fiber the size of a human hair which will transmit lightwaves capable of carrying vast amounts of information.

orbit the path of a satellite around the earth.

PABX (Private Automatic Branch Exchange) a private telephone switching system that provides access to and from the public telephone system.

packet switching a technique of switching digital signals with computers wherein the signal stream is broken into packets and reassembled in the correct sequence at the destination.

PBX (Private Branch Exchange) a private telephone switching system that is not automated.

point-to-point transmission the transmission of electrical signals for reception at specific points as opposed to broadcasting.

public-switched telephone network (PSTN) a name often given to the nation's subscriber telephone services provided by the more than 1500 operating telephone companies, including the operating companies created by the 1984 divestiture of AT&T.

pulse code modulation (PCM) a technique by which a signal is sampled periodically, each sample quantified by means of a digital binary number.

RAM (random-access memory) a RAM provides access to any storage or memory location points directly by means of vertical and horizontal coordinates. It is erasable and reusable.

ROM (read-only memory) a permanently stored memory which is read and not altered in the operation.

slow-scan television a technique for transmitting video signals on a narrowband circuit such as telephone lines which results in a picture changing every few seconds.

software the written instructions which direct a computer program. (See hardware, firmware.)

subcarrier an additional signal imposed on the broadcaster's carrier. This "subchannel" carries information, sound, or data different from that carried by the carrier and requires a special receiver.

tariff the published rate for service, equipment, or facility established by the communication common carrier.

teleconferencing the simultaneous visual and/or sound interconnection that allows individuals in two or more locations to see and talk to one another in a long-distance conference arrangement.

telemetry communication of information generated by measuring devices.

teleprocessing   data processing wherein the data manipulation is performed at a processor electrically connected to but remote from where the data is entered or used.
teletext   the generic name for a set of systems which transmits alphanumeric and simple graphical information over the broadcast (or over a one-way cable) signal using spare-line capacity in the signal for display on a suitably modified TV receiver.
Telex   a dial-up telegraph service.
terminal   a point at which a communication can either enter or leave a communications network.
transponder   the electronic circuits of a satellite which receive a signal from the transmitting earth station, amplify it, and transmit it to earth at a different frequency.
trunk   a main cable running from the headend to a local node which then connects to a drop running to a home in a cable television system. A main circuit connected to local central offices with regional or intercity switches in telephone systems.
TVRO   television receive-only earth station. (See earth station.)
"twisted pair"   the term given to the two wires that connect local telephone circuits to the telephone central office.
UHF (ultra high frequency)   the frequency band from 300 to 3000 megahertz.
uplink   the communications link from the transmitting earth station to the satellite.
VHF (Very High Frequency)   the frequency band from 30 to 300 megahertz.
videotex   the generic name for a set of systems which transmit alphanumeric and simple graphical information over the ordinary telephone line for display on a suitably modified TV set at the request of a user equipped with a numeric keypad. An interactive cable television system also can offer videotex services.
VLSI (very large scale integration)   single integrated circuits which contain more than 100,000 logic gates on one microchip.
WATS (Wide Area Telephone Service)   a service offered by telephone companies in the U.S. which permits customers to make dial calls to telephones on a specific area for a flat monthly charge or to receive calls "collect" at a flat monthly charge.

REFERENCES

DATAPRO Research Corporation, *Communications Solutions: Glossary*. June 1979.
Pennsylvania Public Television Network Commission, *Public Telecommunications in Pennsylvania*. Commonwealth of Pennsylvania, May 1982.
Williams, Frederick and Herbert S. Dordick, *The Executive's Guide to Information Technology*, New York: John Wiley & Sons, Inc., 1983.

INDEX

Active information seeking, 261
Advanced Research Projects Agency (ARPA), 278
Agriculture, workers in:
 Japan, 12
 United States, 9
Alphageometric approach, 253, 254
Alphamosaic approach, 253, 254
Amplifiers, 211, 216
 radio, 198
 satellite, 99
Amplitude envelope, 41
Amplitude modulated (AM) signals, 32, 34, 41-44, 45, 128, 194, 195, 200, 207, 224
Analog systems, 39, 41, 128, 142, 161, 179, 211, 227, 231
AND gate, 64, 65
Answering devices, 160–161
Antennae, 88–92
Appointment in Samarra, 299
ASCII (American Standard Code for Information Interchange), 172
ASEAN, 106
Asynchronous communications, 171, 172, 173

AT&T 3, 19, 145, 180, 225
 See also Bell System
Audio conferencing, 157–158
Audio-visual communications, 158–159, 160
Australian Broadcasting Company, 102
Austrian National Broadcasting Organization, 77

Babbage, Charles, 5, 270
Bandwidth, 37, 38, 39–41, 42, 43, 97, 102, 146, 151, 159, 181, 203, 205, 228, 267, 276, 281
 of AM radio, 207
 of cable television, 210
 of color television, 40
 cost of, 45
 efficient user of, 44, 94, 95, 99, 103, 116, 130, 141, 189, 191, 226, 234, 250
 of hi-fi music, 40
 of mobile radio, 225
 and optical fiber cable, 105, 106, 143
 and SSB transmission, 200
 of telegraph, 40, 169
 of telephone, 40

317

of television, 96, 205, 249
of transponders, 93
of voice, 96, 161
Bardeen, John, 60
Batch processing, 272
Baud, correspondence with hertz, 171
Baudot code, 171
Bell, Daniel, 6, 126, 223, 267, 302, 303
Bell Laboratories, 59, 60, 202
Bell System, 99, 117, 145, 177, 180
  See also AT&T
Bernstein, Leonard, 1
Binary digit (See Bit)
Bit, 35, 169, 171, 172, 173
Blanking interval, vertical, 204, 205, 250, 251, 252, 254, 255
Boole, George, 5, 63, 64
Bowers, Raymond, 235
Brandeis, Louis D., 291
Brattain, Walter, 60
Braudel, Fernand, 5
Braun, Carl, 202
British Broadcasting Corporation (BBC), 202, 251
British Independent Broadcasting Authority, 251
British Post and Telegraph, 19
British Post Office, 257
Broadcast satellite (see Satellites)
Broadcast systems, 30, 41, 113, 187–220, 284
  locating new space in, 199–201
  radio, 197–199, 202, 210
  television, 201–206
Business applications, network information industry, 286
Bytes, 171

Cable radio, 207
Cable spectrum, 209-212
Cable systems, 187-220
Cable television systems, 19, 101, 113, 146, 207-216, 284, 287
Cabletext, 255-256
  tree topology used in, 110, 111, 212
Camera tube, iconoscope, 202
Capacitance, 28
Capacitors, 65
Carty, John J., 126
Cathode ray tube, 202
Cat's whiskers, and semiconductors, 57-59
CCITT (International Telegraph and Telephone Consultative Committee), 182, 183

CEEFAX, 251
Cellular radio, 18, 233-239, 241
Centralized computer communications network, 273, 274
Centrally switched network, 26, 27
Channel(s):
  capacity, 37-38, 96, 143
  communicating, 25
  sharing, 173-175
Chesterton, Gilbert, 299
Chips (see Semiconductor chips)
Circuit design, 63-64
Circuit-switched system, 114
Citizens band service, 232-233
Clarke, Arthur C., 75, 76, 80, 83, 88
Coaxial cable, 207, 208, 209
Color television, 202
  bandwidth of, 40
  See also Television
Common channel interoffice signaling (CCIS), 154-155, 239
Common control switching, 131, 134-136, 149, 156, 163, 231, 236
Communication Act (1934), 145, 218
Communication, speed of, 29
Communication systems, and networks, 24-27, 38
Communications satellites (see Satellites)
Composite video signal, 203, 204, 205
Computers, 5
  as a communications tool, 19, 66-68, 266-282
  link with telephone, 166-183
  mainframe, 270-271
  and telecommunications, 17
Connectivity, of a satellite, 97, 189
Consumer applications, network information industry, 286
Contention, line control, 280
Control signaling, telephone system, 154-155, 156, 163
Converters, cable television system, 213, 214
Cost effectiveness, of communication channel, 29
Crossbar switching, 132-134

Data communications, 21, 174
Data processing costs, 275
Data transport network, 175
Data under voice, 175
de Sola Pool, Ithiel, 262, 263
de Sola Price, Derek, 268
de Tocqueville, Alexis, 264

Decentralized computer communications network, 273, 274
Decoder, 25, 33
  teletext systems, 254
Demodulation, 41, 197
Dialing, subscriber loop, 128
Digital transmission system, 38, 97, 139, 142, 145, 153, 179, 227, 267, 281
  and stored program control, 136-137
  and telegraph, 30, 39, 63, 159, 169
  and telephone network, 176-183
Diodes, 65
  as semiconductor rectifier, 59
  as switch, 60
Direct broadcast satellites (DBS), 101-103, 217
Disk scanning system, spinning, 201-202
Distortion, 28, 30
Distributed data processing network, 273, 276-279, 281
Distribution system, cable television system, 212
Dordick, Herbert S., 301
Downconverter, 205
Downlink, frequency range, 93
Drucker, Peter, 1

Early Bird satellite, 82
Earth stations, 94-100
  television receiver-only (TVRO), 94, 98-100, 101, 102, 103, 216, 217
  transmitting, 95-98
Echoes, 139-141
Educational applications, network information industry, 286
Educational television, 205, 216
Einstein, Albert, 5
EIRP (effective isotropic radiated power), 95, 98
Electrical noise (see Noise)
Electromagnetic field, 31
Electromagnetic radiation, 189
Electromagnetic spectrum, 18, 189-191, 192, 200, 220, 227, 276
Electromagnetism, used in telecommunication systems, 27
Electronic convenience services, 156
Electronic blackboard, 168
Electronic mail, 70, 163, 272
Electronic publishing, 18, 247-263
Electronic switching, 134
Electrostatic fields, 31
Ellul, Jacques, 303
Encoder, information, 25

End-to-end digital system, 178-179
Engels' ratio, 13
ENIAC (Electronic Numerical Integrator and Calculator), 270, 271
EPROM (erasable-programmable-read-only memories), 68
Ergonomics, 118
Escape speed, satellites, 79
European Space Agency, 76
Event clocks, 172
Exchange, telephone, 150-151, 152
Expenditures, of information industry, 10-11

Faraday, Michael, 5, 31, 58
Fiber optics (see Optical fiber transmission)
Fields, electrically charged, 31
Filtering, of noise, 33
FM (see Frequency modulated signals)
Fourier, Joseph, 39
Framing process, 172
Frequency, 39, 41
  and time, 41, 42, 43
Frequency division multiplexing (FDM), 46-47, 96, 97, 101, 141, 142, 153, 154, 178, 191, 207, 227, 228, 235
Frequency modulated (FM) signals, 34, 44-45, 96, 159, 200, 233
Frequency modulation/frequency division multiple access (FM/FDMA), 96
Frequency shift keying (FSK), 34

Galaxy satellites, 93, 99, 101
Gates, 67
  AND, 64, 65
  OR, 64
  transistor as, 63
Geosynchronous orbit, satellites, 79, 80
Gershuny, Jonathan, 303, 305
Gigahertz, 40, 267
Goldmark-CBS system, color television, 202
Government applications, network information industry, 286
Graphic techniques, electronic publishing, 252, 253
Greene Decision (1983), 145

Hardware, substituting for software, 68-70
Harvard Law Review, 291
Headend, cable television system, 212

Hertz, Heinrich, 5, 40, 190
Hertz, 40
  correspondence with baud, 171
Heterodyne, carrier signal, 198
High fidelity music, bandwidth of, 40
Hirschman, Albert O., 305
Horizontal synch pulser, 249
Human factor analysis, 118
Hybrid, 139, 140
  amplifiers, 211, 216

IBM, 136
Iconoscope camera tube, 202
Inductors, 65
Information:
  activities, 8-9
  consumption, 15, 16
  encoder, 25
  equality, 292-293
  explanation of, 35-37
  industry, 10-11
  measuring, 37
  ratio, 13, 14
  seeking, 261
  societies:
    emerging, 3-22
    idea of, 6-7
    and network marketplace, 16-18
    and quality of life, 13-16
    rank of nations as, 14
    speculations about, 299-308
    workforce and, 8-13
  source, 25
  supply, 15
  workers, 9, 12
Institutional cable, 213
Instructional television fixed services
    (ITFS), 102, 205, 216
Integrated circuit, 60, 65-66, 211
Integrated services digital networks
    (ISDN), 137, 178-179, 180, 181, 182,
    183, 279
Intelligent telephones, 70, 148-164, 279,
    289, 293
INTELSAT, 78, 88, 89, 90, 93, 94, 96, 98,
    105, 106
INTELSAT V, 96, 97
INTELSAT VI, 97
Intermediate products, 9
International Telecommunications
    Union (ITU), 93, 195
Interpulse interference, 173

Japan:
  competition in telecommunications,
    20
  consumption of information, 16
  information society concept, 6-7, 300
  supply of information in, 15
  workforce of, 12
Japanese Computer Usage Development
    Institute, 300
Josephson junction, 73
Journal des Savants, 268

Kelley, Mervyn, 59, 60, 71
Kilohertz, 40

Large scale integration (LSI), 72, 271
Leydon Jar, 58, 67
Line-haul transmission, 128, 137-138, 141,
    143, 147, 177, 178
Line-of-sight, 191, 193
Line-switched system, 114
Little Science, Big Science, 268
Loading coils, 168, 169
Local area network (LAN), 277
Local loop, telephone system, 127, 128-
    130, 137, 145, 146, 168, 174
Loud-speaker systems, 41
Lower sidebands, 43
Low-noise amplifier (LNA), 99

Machlup, Fritz, 6, 7, 8, 288, 302
McLuhan, Marshall, 1
Magnetic storage, 66, 67
Magnetostatic fields, 31
Mainframe, 270, 271
Management information systems, 278
Manufacturing workers:
  Japan, 12
  United States, 9
Master antenna, 207
Masuda, Yoneji, 300
Maxwell, Clerk, 5, 40, 190, 200, 201
Medical applications, network informa-
    tion industry, 286
Megahertz, 40
Memory, 66-68, 71
  cost of, 72
Message-switched systems, 114, 272, 277
Microwave services, 18
Minicomputers, 271
MIT, 15

Mixing, radio broadcasting, 197
Mobile communications, economics of, 239-240
Mobile radio, 149, 195, 222-243, 284
Modems, 170
Modulated carrier, 197
Modulation, 32-35, 41, 197
Morse, Samuel, 5, 159
Morse Code, 27, 31, 32, 33, 35, 169, 170
Motion picture industry, and cable television, 214
Multichannel multipoint distribution services (MMDS), 216
Multichannel radio telephone services, 231-232
Multiplexing, 46, 70, 138, 141, 150, 153, 171
 See also Frequency division multiplexing; Time division multiplexing
Multipoint distribution services (MDS), 102, 216

Narrowcasting, 185, 188, 212
NASA ATS-6 satellite, 91
Network information services, 17, 38, 115, 283-295
 economic advantages of, 287-289, 293-294
 elements of, 284-287
 prerequisites for, 289-293
Network marketplace, and information societies, 16-18, 272, 285
Networks, 21, 108-120, 281-282
 architecture of, 108-113, 281
 centralized computer communications, 273, 274
 centrally switched, 26, 27
 and communication systems, 24-27, 38
 point-to-point, 26
 switched and non-switched, 113-115, 151, 228
New York World, 268
Newspapers, electronic, 253-255
Newton, Isaac, 5, 78, 80, 83, 84
Nipkow, Paul, 201
Noise, 28, 29, 30, 32, 33, 38, 100, 138, 173, 194, 198
 and bandwidth, 96
Non-blocking condition, crossbar switch, 134, 153
Non-switched networks, 113-115, 151, 228

N-type semiconductor, 60, 61, 62

Off-hook, subscriber loop, 128
O'Hara, John, 299
One-way paging systems, 227-229, 234
On-Off Keying (OOK), 32, 34, 170
Optical fiber transmission, 105, 106, 143, 145, 156, 189, 267, 306
OR gate, 64
ORACLE (Optimal Reception of Announcements by Coded Line Electronics), 257
Orbital Test Satellite, 76
Orbital velocity, satellites, 79
Orbits, satellite, 79, 80,
Oscillating current, 200
Oscillator, 190

PABX (Private Automatic Branch Exchange), 70, 147, 162
Packet radio, 234
Packet switching, 115, 281
Paging systems, one-way, 227-229, 234
Palapa Satellite, 77, 78, 106
Panther Valley CATV System, 209, 212
Parity check, 173
Passive information seeking, 261
Pay television, 188, 212, 214, 287
Payne Committee, 206
Persistence of vision, 201
Personal computer, 271-272, 273
Phase shift keying (PSK), 34
Philosophical Transactions of the Royal Society, 268
Photolithographic mask, 66, 67
Photo-optical effect, 58
Point-to-many-points communications, 25
Point-to-point communications, 25, 27, 151
Point-to-point network, 26, 130
Polarization, 93, 94
Polling, line control, 281
Porat, Marc, 8, 9, 12, 302
Prestel, 257
Primary information activities, 8
Privacy, network information services industry, 290-292
Privacy Protection Study Commission, 291
Production and Distribution of Knowledge in the United States, 6

INDEX **322**

Programs, computer, 70
PROM (Programmable-read-only-memory), 67
Protocols, 112-113, 117, 171-173, 280, 281
Prove-in/prove-out points, 141, 142
P-type semiconductor, 61, 62
Public switched telephone network (PSTN), 150, 152, 163, 279
  hierarchy of, 151-153
Publishing (see Electronic publishing)
Pulse amplitude modulation (PAM)), 49-52
Pulse code modulation (PCM), 52-53

Quantize, PAM signals, 52

Radio, 18, 19, 189, 218, 223
  broadcasting, 197-199, 202, 210
  spectrum, 191-197
  See also Mobile radio
Radio Act, 218
Radio dispatch systems, 230-231, 234
Radio frequency oscillator, 197, 200
Radio telephone services, multichannel, 231-232
RAM (random-access memory), 66-68, 69
RCA, 202, 251
Receiver, 25
Recipient, 25
Rectifier, semiconductor, 59
Remote data processing, 272
Repeaters (see Transponders)
Resistance, 28
Resistors, 65
Revenues, of information industry, 10-11
Rice, Ronald, 301
Ring network, 113
Ringing, subscriber loop, 128
Ripple current, 283, 284
ROM (read-only memories), 66-68
Rotary press, 267

Sampling, 48-49, 231
  for pulse amplitude modulation, 50
Sampling Theorem (Hartley and Shannon), 51
Satcom satellites, 99, 101
Satellite Business Systems (SBS), 94
Satellite master antenna television (SMATV), 217
Satellites, 21, 75-107, 147, 216, 227, 234, 239, 267, 284
  communications, orbital locations, 87
  cost of, 100

direct broadcast, 101-103, 217
idea of, 78-80
lifetime of, 100
and mobile communications, 241, 242
space segment of, 82-94
synchronous, equatorial parking orbit, 85, 86
technology of, 80-82
uses of, 103-105
Sawtooth modulating signals, 44, 45
SBS (Satellite Business Systems), 99
Scanning pattern television, 203, 204
Scientific American, 156
Scientific information, growth of, 268
Scrambling, cable television system, 214
Secondary information activities, 8
Selenium, 201
Semiconductor chips, 21, 47, 57-73, 150, 183, 189, 211, 229, 234, 271
  cost of, 71, 72
Sequential switching system, 131
Service tiers, cable television system, 213
Service, workers in:
  Japan, 12
  United States, 9
Shannon, Claude, 5, 36, 38, 51
Shannon-Weaver linear communication model, 80-81, 99
Shifting Involvements, 305
Shockley, William, 60
Sidebands, 44
  lower, 43
  upper, 43
Signals:
  Distortion of, 28
  loss of, 138-139
  received, 27, 28, 29, 30
  transmission with noise, 33
  transmitted, 27, 28, 29, 30
  voice, 40
Signal-to-noise ratios, 29, 37, 96, 139, 179, 233
Silicon, 60, 61, 65, 73
Simon, Herbert, 300
Simplex lines, 171, 230
Sine wave, 31, 190
Single sideband (SSB) radio system, 199, 200
Sinusoidal signal, 31
SITA (Societe Internationale de Tele-communications Aeronautiques), 106, 278
Skiatron, 215
Skin effect, 207, 208

Slow-scan television, 159, 168
Smith, Cyril Stanley, 267
Software, substituting for hardware, 68-70
Solar cells, 82
Source, information, 25
Source, The, 173
Space segment, of satellite system, 82-94
  antennae, 88-92
  control and operational telemetry, 84
  cost of, 100
  positioning and orientation system, 83-84
  power source, 82-83
  transponders, 92-94, 99, 103
Speech, transmitting, 38-39
Speed, of communication, 29-30
Stable orbit, satellite, 79
Star topology, telephone system, 109, 150, 273
Step-by-step switching, 132, 136
Stored program control systems (SPC switch), 136
  and digital transmission, 136-137
Structures of Everyday Life, The, 5
Subscriber drop, cable television system, 213
Subscriber loop, telephone system, 127, 128-130, 154, 169
Subscription television (STV), 188, 215
Superheterodyne receiver, 199, 210
Surgeon General's Report on Television and Violence, 206
SWIFT network (Society for Worldwide Interbank Financial Telecommunications), 106, 278
Switch:
  diode as, 60
  gate as, 64
Switched networks, 113-115, 151
Switching, telephone, 60, 127, 130-136, 147, 156, 176-177, 178, 213
Synchronizing pulses, 249
Synchronous satellite, equatorial parking orbit, 85, 86
Synchronous transmission, 172

Tarleton, Robert, 208
Technical concepts, 24-53
Technologies of Freedom, 262
Telecommunication systems, 27-30
Telecommunications technology:
  major figures in development of, 4-5
  and transportation, 156-158

Teleconferencing, 70, 157-158, 160
Telegraph, 5, 19, 21, 34, 103, 104, 114, 171, 222, 223
  bandwidth of, 40
  as digital communications system, 30, 39, 63, 159, 169
  signal, mechanics of sending, 32
  use of electricity in, 27
Telemeter, 215
Telemetering, 84
Telephone services, multichannel radio, 231-232
Telephone system, 5, 18, 21, 100-101, 104, 116, 201, 222, 223, 284
  architecture of, 150-153
  as analog communications system, 39
  bandwidth of, 40
  channels, 196
  digital, 176-183
  income from, 103
  intelligent, 70, 148-164, 279, 289, 293
  link to computer, 166-183
  network architecture of, 109, 110
  number of, 103, 126
  politics of, 179-183
  subscribership (1945-1975), 19
  switching, 60, 113, 114
  technology, 125-147
  three-chip, 68, 69
Teleprinter, 171
Teleprocessing network, 272, 280
Teletext, 18, 247-263
Teletype, 170
Television, 18, 19, 116, 218, 220
  bandwidth of, 92, 205
  broadcasting, 201-206
  slow-scan, 159, 168
  See also Color television
Television receiver-only (TVRO) earth stations, 98-100, 216, 217
Telex, 104, 170
Terminals, 21, 70, 109-120
Threshold, of detection, 28, 29
Time, and frequency, 41, 42, 43
Time division multiplexing (TDM), 47-48, 51, 97, 99, 115, 141, 142, 153, 161, 176, 177, 178, 191, 234, 267, 275, 280, 281
Time division multiple access (TDMA), 97, 104
Time-sharing, teleprocessing network, 272, 273
Time-space switching, 176-177, 183
Transistors, 60, 65, 67
  principles of, 62-63

Transmission economics, 141-144
Transponders, 92-94, 99, 103
Transportation, and telecommunications, 156-158
Traveling wave tube (TWT), 92-93
Treatise on the Law of Tort, (1879), 290
Tree-topology, cable television system, 110, 111, 212
Trunk system (see Line-haul transmission)
Tuned radio frequency (TRF), receivers, 198
Turner, Frederick Jackson, 222
TVRO (television-receiver-only) earth stations, 94, 98-100, 101, 102, 216, 217
   cost of, 103

Ultra high frequency range (UHF), 193, 210, 211, 224, 225
Uplink, frequency range, 93
Upper sidebands, 43
U. S. Department of Commerce, 195
User friendliness, of computer equipment, 119
U. S. Federal Communications Commission (FCC), 195, 196, 200, 205, 214, 215, 216, 224, 225, 227, 233, 251
U. S. Supreme Court, 196

Vacuum tube, 65
Vail, Theodore, 145
Value-added networks, 117
Vertical blanking interval, 204, 205, 250, 251, 252, 254, 255
Vertical retrace, 250
Very high frequency range (VHF), 193, 210, 211, 224, 235
Very large scale integration (VLSI), 72, 271
Video display terminal (VDT), 118, 119
Video signal, composite, 203, 204, 205
Videotex, 247-263, 289
Viewdata, 257
Vision, persistence of, 201
Voice channels, 173, 174
Voice mail system, 163
Voice network, data on, 168-175
Voice signals, 40, 41, 162
   bandwidth of, 96, 161
   frequency multiplexing, 46
Voice transmission, subscriber loop, 128
Von Neumann, John, 5, 270

Warren, Samuel, 291
Wavelengths, 190, 191
Welles, Orson, 185
Wells, H.G., 156
WESTAR, 82, 96, 101
Western Electric Company, 117, 136, 145
Williams, Frederick, 301
Wireless cable, 215-216
Wireless communications, 30-32
Wireless Ship Act (1910), 218
Workforce, and the information society, 8-13

Zworykin, Vladimir, 202